Dyscalculia Demystified
Practical Tools for the General Education Classroom

Dawn Pagliaro-Newman

PolyMath Publishing

Osage, IA

2026

PolyMath Publishing
403 Chase Street
Osage, IA 50461

ISBN 978-1-7327601-9-6

Book Cover Design by Melli M Designs
Interior Design by Evan Scott
First printing edition 2026

www.dawnpagliaronewman.com

Praise for *Dyscalculia Demystified*

What excites me about this book is that it does not push more of the same or lean on gimmicks to "fix" math struggles. Dawn grounds her ideas in both experience and science, and that matters right now, as math achievement is declining nationwide.
— **Jon Harper, 4th & 5th Grade Math Teacher, Host of *Teachers' Aid* on the BAM Radio Network**

This book is practical, offering concrete steps for addressing specific math challenges while reminding teachers to base decisions on what they observe in students. The section on partnering
with families is especially powerful, thoughtful, thorough, and deeply needed. Dawn's perspective as both a teacher and a parent makes this guidance feel genuine and grounded. This is a valuable resource that will help so many educators.
— **Allison McDonald, M.S., Founder of *No Time For Flashcards***

This book takes an asset-based approach to understanding dyscalculia, highlighting the unique strengths and problem-solving abilities of each learner. With clear, actionable steps, it empowers educators and families to support students in reaching their fullest potential.
— **Jessie Startup-DeNat, Math Instructional Lead, District 15, NYC Public Schools**

For Olivia,

Never forget: You change the world by being fully,
bravely, and beautifully yourself.

Table Of Contents

Forward by Dr. Honora Wall

Teaching is a complex, multifaceted activity that requires dedication, training, and expertise. Teaching students who struggle also requires one to be a detective. Why is this student still struggling after we've used the right instruction and great interventions? How can they seem to forget everything they just did a few days ago? What in the world are we supposed to do to help this child succeed? Sometimes the answers are easy: add some games to class time, assign extra practice, send home ideas for things to do as a family. Sometimes the answers are hard to find. The standard approach doesn't work. We know a student is bright and capable, but we don't know why their struggles persist. This is usually our first indicator that a student might have a learning disability. The math-related learning disability is called *dyscalculia*. Those of us who study dyscalculia are master investigators; this book will help you join our ranks.

Dawn Pagliaro-Newman began her investigative journey in order to support her students and her daughter. During her travels, she found me; we became acquaintances, then colleagues, and then friends. Dawn didn't stop there, of course. She kept reading, researching, and trying different techniques in her classroom. She attended and presented at conferences. She wrote articles on dyscalculia. She started a support group, *OverThink Tank*, for fellow teachers beginning their journey into the world of math learning disabilities. And now, she has written an informative book on dyscalculia that is relevant to all classroom teachers.

Of course, there's more to Dawn than just dyscalculia. She brings more than two decades of teaching experience to this book, along with both a Math for America and New York State teaching Fellowship. She leads teacher workshops in her local district. She appears on podcasts, and is a devoted wife and mother. She is a dedicated, exceptional teacher.

This book is written for general education classroom teachers, putting dyscalculia into the ecosystem of 30-ish students, 36 weeks, numerous holidays, and high-stakes state tests. Readers will learn about the math disability, other common math struggles, and ways to support students within their regular math class. It also helps teachers become better at recognizing when a productive struggle is actually a more serious issue. It helps teachers know how to tease out strengths and weaknesses. And it helps teachers balance these tasks within their already crowded workday. Keep a highlighter and sticky notes nearby as you explore these pages! This is a resource you will come back to, over and over again.

Honora Wall
January 2026

Acknowledgements

This book was not created alone. It is the result of encouragement, community and love.

To my husband: Thank you for believing in me, for giving me the confidence and space to create, and for supporting me as this dream took shape.

To my daughter: You are my inspiration and guiding light. This work is dedicated to you.

To my parents: Thank you for being my biggest cheerleaders and for instilling in me the value of persistence, curiosity, and work worth doing. To my brother, Steven, and my sister, Laura: Thank you for your steady support and guidance. To Jonathan Berger: Thank you for making sure this dream was built on solid ground.

To Jessica Ceballo: Thank you for seeing this before I could.

To Melissa Singer, whose coaching, collaboration, and unwavering belief in what is possible sustained me and shaped my practice.

To the PS 130 family, my work home: Thank you for giving me the freedom to question, explore, and try. Thank you for working side by side with me and shaping the educator I am.

To Nicole Lopez and Patti Ann Vazquez: Thank you for your guidance, encouragement, and clarity when I needed it most.

To Math For America: Thank you for the community you have cultivated and your shared commitment to excellent teaching. To Edutopia: Thank you for encouraging me to write more and to share what I was learning.

To Honora Wall: Thank you for giving a megaphone to those with unheard voices and for taking a chance on me.

To the OverThink Tank, Dawn Cohen Frank, Maureen Stewart, Jennifer Levy, Mufridah Nolan, and Michele Du Maresq: Thank you for stretching my thinking and helping me shift how I see and support the students I serve.

To my generous beta readers: Angie Anderton, Kendra Din, Andrea Kung, Alexander Lord, Katherine Maschmeyer, Allison McDonald, M.S, and Jessica Smith: Thank you for your time, honesty, and care. Your insights strengthened this work.

To Evan Scott and Melanie Moor: Thank you for bringing this book to life visually. Your creativity honors the heart of this book.

And most of all, to my students: twenty-five years of Leo's and Sofia's who continue to teach me, humble me, and inspire me.

Thank you for trusting me, challenging me, and reminding me what matters most.

Introduction

The bell rings like a starting pistol before you even have a chance to sip your coffee. Thirty students file into a room designed for 20, bringing with them the movement, questions, and needs that swirl into the usual morning flurry. Someone's lost their homework. Someone needs help unzipping their coat. There is a fight brewing over the blue marker. One student needs a behavior check-in. And what is that smell? All the while, your inbox is full of reminders - benchmark testing, professional development sign-ups, two schedule changes, and that one parent "just checking in" again.

This is the reality of the general education classroom: a place where teachers are expected to meet every need, reach every learner, and make it look effortless. But beneath the surface lies a quiet truth: some students are slipping through the cracks. It's not because teachers don't care. Through it all, we do what we do best - adapt, support, and juggle. Because beneath the noise and never-ending to-do lists, we know that every student is worth the effort.

Amid all the bustle, there are students whose needs are not as blatant as a backpack or a blue marker. They are the ones who put in so much effort, only to make the same mistakes over and over again. The ones who seem to master a skill on Tuesday, only to stare at the paper on Thursday as if they've never seen it before. Or maybe they are the ones who suddenly need to use the bathroom, or visit the nurse, as soon as numbers appear. They're not defiant. They're not disengaged. And despite our best efforts, they just don't make the progress we hope for.

At first, we try small groups, manipulatives, and reteaching the concept one more time. Maybe we see a few small gains. But the curriculum moves on, and no matter how many times we document, adjust, and reteach - no matter what tool we have in our arsenal - progress just doesn't come.

Sometimes, it's not a gap in instruction, or a missing prerequisite skill. Sometimes, what we are seeing is something else entirely: a learning difference that is rarely discussed and often overlooked - dyscalculia.

Dyscalculia is a specific learning disability that makes it hard to understand and work with numbers. Students with dyscalculia may struggle to make sense of quantities, remember math facts, perform mental calculations, or connect one math concept to another. Word problems can feel overwhelming. Information doesn't seem to stick, despite their best efforts. Like dyslexia, or ADHD, dyscalculia is something people are born with. Like other learning differences, dyscalculia can present on a spectrum. No two students experience it exactly the same way which means each learner brings a unique set of strengths and challenges.

The good news is that there are many ways to support these learners. Dyscalculia isn't a mystery to solve, but a learning difference that can be understood and planned for. And that never-ending to-do list - the one that is probably growing even as you read? This book isn't here to add more tasks that may or may not get crossed off before June. Instead, it offers a new lens - a way of thinking about the planning and teaching you are already doing. With the right tools, strategies, and mindset, we can help these students build confidence, make meaningful progress, and feel like they truly belong in the math classroom.

I don't have all the answers. But I have spent years asking the same questions many teachers are asking now. I never planned to spend my evenings reading research articles, attending webinars, and digging through books on learning differences. Like most general education teachers, I wanted to help all my students reach high standards. And by most measures, I did.

I began teaching in 2001 in a Title I elementary school in Brooklyn, New York. At the time, No Child Left Behind had reshaped education. High stakes standardized tests were tied to student promotion and teacher evaluations. I worked hard to make sure my students succeeded, and the results reflected that. My principal noticed. I was asked to serve on committees, to open my classroom for visiting teachers from across the district, and to lead professional development at the school and district levels. Eventually, I earned my master's degree in special education.

Sure, there were always a few students each year who didn't make the progress I hoped for. I chalked it up to all the usual reasons - attendance issues, home challenges, or just something I couldn't quite pinpoint. I started to think of it as part of the job. Unfortunate, but unavoidable.

And then I had my daughter.

She started school curious and eager. She loved music and listening to stories. But when it came time for math or reading, she became avoidant. At home, seemingly simple homework assignments turned into impossible tasks. Every strategy we tried -

strategies I had used successfully with my own students - did nothing to help the screaming six-year-old who just wanted to escape to her room to play with dolls. When avoidance didn't work, she got angry. Daily meltdowns led to her being removed from the classroom. Our kind, happy child was becoming unrecognizable - clearly in crisis.

We knew something was wrong. But even with all my years of training and experience, I couldn't make sense of what was happening. When she was finally diagnosed with dyscalculia I was stunned - not just by the diagnosis itself, but by this new word I had never encountered before. How is it, with two college degrees and all my experience, I had never heard of something that was so clearly affecting not only my daughter, but likely many of the students I taught over the years?

That question lit a fire in me. I immersed myself in research, reading everything I could find. And as I started sharing what I was learning with colleagues, I realized something surprising - many of them had never heard of dyscalculia either. I was determined to change that - not just for my daughter, but for every child like her, sitting in classrooms, quietly (or not so quietly, in her case) wishing they could escape.

I began writing articles for Edutopia, attending and presenting at conferences given by the National Council for Teachers of Mathematics, and interviewed for podcasts with BAM Radio Network. I became the two-time recipient of the Math for America Master Teacher Fellowship. As my journey continued, my goal became clear: increase awareness, develop practical tools, and build a community where educators could learn from each other. That community became *The OverThink Tank*, a space for teachers to reflect, share, and support each other in helping students with dyscalculia. Over time, those conversations, lessons, and shared stories led to this book.

This book also draws on decades of research in math education, special education, and cognitive science. You will see the influence of thinkers like Brian Butterworth, who helped define dyscalculia; Honora Wall, whose expertise blends the roles of educator and advocate; and guidance from organizations such as the National Research Council and the What Works Clearinghouse. You'll also find practical applications of frameworks such as the CRA (Concrete-Representational-Abstract) sequence - tools that can be mindfully woven into the planning you are already doing to support students with dyscalculia more effectively, regardless of the mandated curriculum or grade level. The goal isn't to add more to your plate, but to help you view your instruction through a more inclusive, strength-based lens: to make more flexible, in-the-moment choices grounded in evidence-based practices, and to know when it is time to advocate for additional support services. This book is designed to walk with you - not ahead of you - through the process of understanding and supporting students with dyscalculia.

Chapter 1: What Is Dyscalculia, Really?

Before we can plan effective instruction for students with dyscalculia, we first need to understand what it is, and what it isn't. We'll compare two leading definitions - from the DSM5 and the SpLD Assessment Standards Committee in the United Kingdom, which will help us identify and dispel common misconceptions. This will make it easier for us to distinguish behaviors that may be part of the disability and what may be attributed to other factors.

Chapter 2: Dyscalculia In The Classroom

Next, we will take a closer look at how dyscalculia can manifest in real-world classrooms by exploring four common subtypes. Understanding these subtypes can help us anticipate the kinds of mistakes a student is likely to make, which is a crucial first step in planning effective instruction. To ground this in practice, we'll meet Leo and Sofia - two students who may remind you of some of your own.

Chapter 3: Identifying Strengths

In this chapter we will shift our lens from focusing on "what's missing" to recognizing "what's working". We'll explore the *Strands of Mathematical Proficiency* and consider how these strengths - often overlooked in students with dyscalculia - can be leveraged as powerful entry points for learning new skills.

Chapter 4: Where Are They Now?

The final step in creating a student profile is determining what the student already understands. In this chapter we will examine Leo and Sophia's work through the lens of the CRA

(Concrete-Representational-Abstract) Instructional Sequence, exploring ways to adapt the sequence to meet the individualized needs of learners with dyscalculia. This process allows us to design meaningful, achievable learning goals that align with the curriculum already in use - supporting students without requiring a complete instructional overhaul.

Chapter 5: Planning For Instruction

Armed with insights into a student's subtype, strand strengths, CRA phase, and clearly defined learning goals rooted in the current curriculum, we are ready to plan for instruction. Differentiation, accommodations, and modifications are part of our everyday teaching practice. But we are still left with a critical question: How do we teach grade-level standards to students who are two or more years behind - within the realities of a general education classroom? This chapter will explore six high leverage instructional moves that can be used strategically to increase access and support success: mathematical language, manipulatives, worked examples, think aloud, targeted feedback, and games.

Chapter 6: There's No I In Team: Family-School Partnerships

No classroom is an island. It truly takes a village. While phrases like these may sound cliche, they reflect a crucial truth: supporting students with dyscalculia is a collaborative effort. This chapter explores how to build strong partnerships with families, offering practical strategies for communicating and empowering caregivers at home. We'll also discuss when and how to engage your school's support service providers and intervention team, ensuring that students receive consistent coordinated support.

At its core, this book is about community - about making sure that every student, regardless of how they process numbers, knows that they are an important part of the math classroom. My goals are threefold: for you, your students, and myself.

I want you to feel more confident and informed when teaching students with dyscalculia. What you will find here won't ask you to completely overhaul your teaching. It's about fine-tuning practices you already use with intention. You'll gain tools to recognize when additional support services or interventions are needed as well the language to advocate with confidence. My hope is by the end of this book, you feel more equipped, less overwhelmed, and more connected to your students.

All students deserve to experience success in math, But for students with dyscalculia, all too often the focus is on what they can't do. We want them to know that they bring unique gifts and perspectives to the classroom. We want them to feel seen and valued, to develop a positive math identity, and to experience a classroom where their voices are heard.

This book is about sharing what I've learn - what works, what doesn't, and what's worth your time. As educators, our plates are already full. My promise to you is to keep things clear, purposeful, and useful. My goal isn't to add more, but to make teaching math to students with dyscalculia feel more manageable.

The decisions we make shape how students with dyscalculia experience math and how they see themselves as learners. The work isn't always easy, but it's important, and deeply rewarding. And the good thing is, you don't have to do it alone.

References

Butterworth, B. (2019). *Dyscalculia: From science to education.* Routledge.

Harper, J. (Host). (2024, June 17). *Unlocking potential: Innovative teaching strategies for students who struggle with math* [Audio podcast episode]. In *BAM! Radio Network.* BAM Radio Network. https://www.bamradionetwork.com/track/unlocking-potential-innovative-teaching-strategies-forstudents-who-struggle-with-math/

Karagiannakis, G., Baccaglini-Frank, A., & Papadatos, Y. (2014). Mathematical learning difficulties subtypes classification. *Frontiers in Human Neuroscience, 8,* 57. https://doi.org/10.3389/fnhum.2014.00057

National Research Council. (2001). *Adding it up: Helping children learn mathematics* (J. Kilpatrick, J. Swafford, & B. Findell, Eds.). National Academy Press. https://doi.org/10.17226/9822

Pagliaro-Newman, D. (2024, April 25). *Shining A Light On Dyscalculia*. Edutopia. https://www.edutopia.org/article/teaching-students-dyscalculia

Pagliaro-Newman, D. (2025, August 6). *What Does It Mean To Be Good At Math?*. Edutopia. https://www.edutopia.org/article/being-good-math-elementary-students

Pennsylvania Training and Technical Assistance Network. (n.d.). *Concrete-representational-abstract: Instructional sequence for mathematics* [PDF]. PaTTAN. https://www.pattan.net/getmedia/9059e5f07edc-4391-8c8e-ebaf8c3c95d6/CRA_Methods0117

SpLD Assessment Standards Committee. (2025, March 24). *Guidance on the assessment of mathematics difficulties and dyscalculia*. SASC. https://www.sasc.org.uk/news/maths-difficulties-and-dyscalculiaguidance-march-25/

U.S. Department of Education, Institute of Education Sciences, What Works Clearinghouse. (2021). *Assisting students struggling with mathematics: Intervention in the elementary grades* (WWC 2021003). National Center for Education Evaluation and Regional Assistance. https://ies.ed.gov/ncee/wwc/PracticeGuide/21

Wall, H. (2022). *Teaching Students with Dyscalculia*. PolyMath Publishing.

Chapter 1 What Is Dyscalculia, Really?

As educators, we have all encountered students who struggle with math. But what does that actually mean? If we were to sit around a table and each describe a student we have worked with, certain patterns would emerge. We'd hear about students who struggle to retain basic facts, who confuse steps in multi-step procedures, and who seemingly forget concepts that appeared to have been mastered the day before.

Yet, there would be certain behaviors that would be very different. One educator may describe a student who can explain math concepts out loud, but freezes when it's time to write them down. Another might share of a student who excels in geometry, but struggles to count past 30. Someone else may work with a student who is emphatically sure that the sum of 32 and 88 is 1110 and will not be convinced otherwise. These variations matter. They remind us that math difficulties are not one size fits all, and that not every student who struggles with math has dyscalculia.

This is where the danger of misunderstanding comes in. Without a clear framework, we run the risk of overidentifying or overlooking students. We may attribute persistent difficulties to dyscalculia when they stem from something else entirely such as interrupted instruction, language processing differences, executive function challenges, or even trauma. On the other hand, we may dismiss real learning differences as temporary or simply developmental. Although age and maturity play a role in development, "having a late birthday" or "just give it time" can't account for *years* of confusion. This misconception runs the risk of delaying time sensitive, targeted support.

Let's be clear: it's not our job as educators to diagnose dyscalculia. Doing so falls well outside the scope of both our training and job description. Formal diagnosis requires an evaluation from a licensed specialist. But that doesn't mean we don't have an incredibly important role to play. Our observations, documentation, and instructional experiences are often the first clues that further action is needed. As classroom teachers, **we** are the ones who have the opportunity to build a support network around those we serve.

While not all math difficulties are attributed to dyscalculia, chances are your student won't be diagnosed with it even when it is. Dyscalculia research is significantly underfunded, and as a result underdiagnosed especially for those who skillfully mask their struggles with strong verbal skills, high effort, or quiet compliance. Cognitive neuropsychologist Brian Butterworth writes that since 2000, the National Institutes of Health has spent $107.2 million on funding for dyslexia research, but only $2.3 million on dyscalculia. This makes it all the more important that we know what dyscalculia is (and what it isn't) so that we can respond skillfully when we see signs of it in our students.

More Than Just a Struggle: What The DSM-5 Says

In the United States the DSM-5 is used by clinicians and insurance companies, providing a common language around disorders and their diagnosis. The DSM-5 does not use the word dyscalculia. Instead, someone can be diagnosed with what is called Specific Learning Disorder with Impairment in Mathematics. Because Specific Learning Disorder in the DSM-5 includes criteria for reading and writing as well, you will notice numbers missing in the definition below. That is because our focus is only on the criteria pertinent to mathematics.

Specific Learning Disorder

Diagnostic Criteria

1. Difficulties learning and using academic skills, as indicated by the presence of at least one of the following symptoms that have persisted for at least 6 months, despite the provision of interventions that target those difficulties: …

   5. Difficulties mastering number sense, number facts, or calculation (e.g., has poor understanding of numbers, their magnitude, and relationships; counts on fingers to add single-digit numbers instead of recalling the math fact as peers do; gets lost in the midst of arithmetic computation and may switch procedures).

   6. Difficulties with mathematical reasoning (e.g., has severe difficulty applying mathematical concepts, facts, or procedures to solve qualitative problems).

2. The affected academic skills are substantially and quantifiably below those expected for the individual's chronological age, and cause significant interference with academic or occupational performance, or with activities of daily living, as confirmed by individually administered standardized achievement measures and comprehensive clinical assessment. For individuals age 17 years and older, a documented history of impairing learning difficulties may be substituted for the standardized assessment.

3. Learning difficulties are not better accounted for by intellectual disabilities, uncorrected visual or auditory acuity, other mental or neurological disorders, psychosocial adversity, lack of proficiency in the language of academic instruction, or inadequate educational instruction.

Source: DSM-5, 2013

While this definition is written for clinicians, there is much here that is meaningful for educators. The DSM-5 outlines specific patterns many of us recognize in our classrooms such as persistent struggles with number sense, basic calculation, and mathematical reasoning. What's key here is *persistence, severity,* and *resistance to intervention.* These aren't small gaps or mild struggles. These students receive instruction and still their understanding of quantity, procedure, and application remain well behind the expectations of their age.

This framework raises important questions for us as educators:

- Have I observed persistent math difficulties that don't improve with reteaching or targeted small group instruction for **more than** 6 months?

- Does this student rely heavily on strategies (like finger counting) that their peers have moved beyond?

- Do they struggle to make sense of concepts, relationships, and procedures <u>despite</u> repeated exposure?

- Could there be other factors such as limited English proficiency, or inconsistent attendance that could be contributing to what I am seeing?

These types of questions are meant to focus our lens as educators. They guide us to be more precise when we observe, purposeful with our instructional choices, and confident when advocating for our students who struggle with math.

Understanding Dyscalculia Through The SASC Lens

Another important source of information to consider comes from the SpLD Assessment Standards Committee (SASC) in the United Kingdom. The SASC provides professional guidance for assessing and diagnosing specific learning difficulties, including dyscalculia. While the committee operates under the UK educational and healthcare systems, its work has global relevance. SASC guidance is rooted in key, recent research to define what dyscalculia is, what it is not, and what has yet to be defined. This offers precise criteria and research-informed guidelines for understanding and diagnosing the condition. In countries like the United States, where definitions can feel ambiguous, SASC materials serve as a touchstone for educators, psychologists, and researchers seeking clarity and shared language. Their emphasis on early identification, standardized assessment practices, and the distinction between dyscalculia and other math related challenges, help move the field toward greater consistency and equity in support.

The complete definition is written in language geared toward professionals in cognitive science and educational psychology. However, there are several key ideas that are especially relevant for our work as classroom teachers and can become a powerful tool for understanding what our students are experiencing. The more we know, the better we can support them.

SASC Definition of a Specific Learning Difficulty In Mathematics (2025)

Features: A specific learning difficulty in mathematics is a set of processing difficulties that effects the acquisition of arithmetic and other areas of mathematics.

In dyscalculia, the most commonly observed cognitive impairment is a pronounced and persistent difficulty with numerical magnitude processing and understanding that presents in age related difficulties with naming, ordering, and comparing physical quantities and numbers, estimating and place value.

Some individuals may not present with a specific cognitive impairment in numerical magnitude processing but have an equally debilitating specific learning difficulty (SpLD in mathematics) due to other processing difficulties. Difficulties in language, executive function (visuo-spatial working memory, inhibitory control) and visual-spatial processing may also contribute.

Impact: Mathematics is a very varied discipline. Difficulties with learning mathematics may present in specific areas (for example, basic calculation) or across the mathematics studied by the individual in relation to age, standard teaching and instruction, level of other attainments. Across education systems and age groups, difficulties in arithmetic fluency and flexibility and mathematical problem solving are key markers of a SpLD in mathematics. Persistent difficulties in mathematics can have a significant impact on life, learning and work. This may also have a detrimental impact upon an individual's resilience to apply mathematical skills effectively.

Presentation: The presentation and development trajectory of a specific learning difficulty (SpLD) in mathematics depends on the interactions of multiple genetic and environmental influences. It will persist through life but my change in manifestation and severity at different stages. It will persist through life but my change in manifestation and severity at different stages.

A SpLD in mathematics frequently co-occurs with one or more of the following: attention deficit hyperactivity (ADHD), dyslexia, developmental language disorder (DLD) and developmental coordination disorder (DCD).

Maths anxiety commonly co-occurs with a SpLD in mathematics but is not an indicator in itself.

Source: SASC Guidance on Assessment
of Mathematics Difficulties and Dyscalculia, 2025

The SASC makes a distinction between what they call a specific learning difficulty in mathematics and dyscalculia. A big part of that language centers around *cognitive processing* – in other words, the mental systems involved in how we take in, understand, and work with information. According to the SASC, a specific learning difficulty in mathematics can be influenced by a range of factors. But only two are essential for their definition of dyscalculia: numerical magnitude processing and mathematical vocabulary.

Numerical Magnitude Processing

In the past, one of the most commonly accepted indicators of dyscalculia has been difficulty with number sense. But the term number sense is broad and often loosely defined. A cognitive scientist might describe it as an internal system for estimating and comparing quantities, while a second grade teacher might think of it as a student's ability to count on, make ten, or understand place value.

Because of these varying definitions, the SASC uses the term **numerical magnitude** processing to include several concepts that are important to our work as teachers. This includes **non-symbolic and symbolic magnitude processing**, which sounds a bit more technical than simply saying "they're working on subitizing." It also includes concepts that we have all actually heard of such as estimation, counting, and sequencing.

Non-symbolic magnitude processing, also known as **subitizing, is the ability to instantly recognize** how many items are in a small group without needing to count. For example, if I pull three nickels from my pocket, I don't need to count them one by one; I just know there are three. In the early elementary grades, teachers often use dot patterns and ten frames to work on this skill. *Perceptual subitizing* refers to quickly recognizing very small quantities (like 2 or 3), while *conceptual subitizing* involves mentally organizing larger groups into smaller chunks (like seeing 6 as two groups of 3, or 10 as 5 and 5). These visual and intuitive understandings are essential stepping stones to arithmetic, estimation, and number comparison. When students struggle with subitizing, or non-symbolic magnitude processing, even simple numerical relationships can feel blurry or unreliable, making the rest of math more difficult to access.

Closely related to this is **symbolic magnitude processing, the ability to understand and reason** about numbers when they are written as digits or numerals, rather than shown with physical objects. It's what allows a student to recognize that 5 is less than 8 without needing manipulatives or pictures. Symbolic magnitude processing supports important skills like mental math. It's also critical for developing place value understanding, such as knowing that 52 is smaller than 205, not just by looking at the digits, but by understanding what those digits represent in their positions. When students have difficulty with symbolic magnitude processing, they may struggle to order numbers, judge the reasonableness of their answers, or make sense of numerical relationships in written problems. Building strength in this area lays the groundwork for *flexible thinking* and success with more advanced mathematical concepts.

In addition to symbolic and non-symbolic magnitude processing, **numerical magnitude also includes skills like estimation, counting, and sequencing**. These abilities help students make sense of number relationships, judge reasonableness, and organize mathematical thinking. For students with dyscalculia, difficulties in any of these areas may cause foundational cracks that make even basic math feel confusing or unpredictable.

Estimation is the ability to make a reasonable guess about quantity, value, or measurement without needing to calculate an exact answer. This skill allows students to judge whether a solution makes sense or to quickly compare values in everyday situations. For students with dyscalculia, estimation is often weak or underdeveloped. They may rely heavily on exact counting, even when precision isn't necessary, or they might produce wildly inaccurate guesses – like saying there are 100 pencils in a cup with only 20. In the classroom, teachers might notice that these students don't recognize that 98 is close to 100, or that 3+4 shouldn't give them a sum in the 20s. They may also struggle with benchmark numbers (like estimating to the nearest 5 or 10) or judging whether their answer is too high or too low.

Counting seems simple, but it involves several *interconnected* skills: saying number word in the correct order (rote counting), coordinating those words with objects (one-to-one correspondence), understanding that the last number counted represents the total (cardinality), and recognizing that the number of items doesn't change if their order does (order irrelevance). Students with dyscalculia may count inaccurately or inconsistently, skip numbers, double count, or forget where they left off. Others might not understand that "five" means the total set, not just the last item touched. Teachers may notice students who count every item, even when a group is small enough to subitize, or who recount again and again due to lack of confidence in their answers.

Sequencing is the ability to understand and remember the order of numbers, steps, or events in a process. In math, that means more than just reciting numbers in order. It involves recognizing what comes before and after, arranging values from least to greatest, and keeping track of multi-step procedures. For students with dyscalculia, sequencing challenges can show up in *several* ways. In early elementary grades, they may count forward or backward inaccurately, skip numbers, or misidentify what number is missing in a sequence (e.g., 23, ___, 25). They may also mix up the steps of a routine procedure, such as regrouping in addition.

In upper elementary, sequencing issues can interfere with more complex tasks, especially when working with fractions and decimals. A student might not understand that ¼ comes before 1/3 on the number line, or might think that 0.9 is smaller than 0.45 because they are comparing digits rather than place value. When asked to order unit fractions from least to greatest, a student might place ½, 1/3, and ¼ in the wrong order because they don't have a reliable internal framework for sequencing values. You might also see students misapply procedures for multi-digit multiplication or long division because they forgot which step comes next or perform the steps in the wrong order.

Mathematical Vocabulary

Alongside numerical magnitude processing, the second key component of the SASC's definition is difficulty with mathematical vocabulary. This doesn't just mean words like denominator and polygon. **It includes everyday language that takes on a specific meaning in math, such as more, less, between, over or how many more.** These linguistic elements may seem simple on the surface, but for many students, especially those with dyscalculia, act like invisible hurdles. If students don't fully grasp what a word or phrase is asking them to do, it doesn't matter how well they understand the numbers involved.

Consider a word problem that asks, "How many more apples does Miriam have than Jordan?" A student who struggles with the concept of more than might add the number together instead of subtracting, or not know what to do at all. Other students might confuse terms like difference or sum, or misinterpret over as meaning "above" in space rather than as a fraction or division. These language-based misunderstandings can masquerade as calculation errors when, in fact, the problem lies in comprehension.

This difficulty becomes especially noticeable in multi-step problems, interpreting graphs and data sets, or when new terms are introduced quickly. For example, multi-step word problems rely on subtle vocabulary, like combined, altogether, or times as many, that signal which operation to use. Being able to identify and understand these words makes it possible to identify important information, eliminate extraneous details, and decide on a sequence needed to find a solution. Even small shifts in phrasing such as "How many in all?" versus "How many more?" require different reasoning.

Students with dyscalculia may need much more exposure to these words, with explicit instruction that connects language to meaning, not just once, but repeatedly and in multiple formats. For example, teachers might model how to underline key information, build visual word walls, or display sentence frames to support discussion. Over time, linking verbal, visual, and symbolic representations help solidify understanding and reduce cognitive load.

Dispelling Common Misconceptions

When I talk with teachers about dyscalculia, I often hear the same myths repeated. Clearing up these misunderstandings is an important first step to recognizing and supporting students who struggle with math.

One common misconception is that dyscalculia is simply *dyslexia with numbers*. While dyslexia typically involves difficulty learning to read, dyscalculia is rooted in challenges making sense of numbers and mathematical concepts. Students may experience both, but the two are distinct learning differences.

Another misconception is that dyscalculia is rare. In fact, research suggests it is just as common as dyslexia. Brian Butterworth notes that between 50 and 60 percent of individuals with dyslexia also have dyscalculia—meaning that, rather than being unusual, it is likely present in every classroom.

Dyscalculia is not the same as "being bad at math." Butterworth also highlights research showing that even infants demonstrate an innate sense of numerosity. Dyscalculia reflects a specific difficulty in processing numbers, not a lack of intelligence or effort. With the right support, students with dyscalculia can build strategies to access mathematics and succeed.

Not all students with dyscalculia struggle in all areas of math. Students with dyscalculia display strengths just as any other student would. Some students demonstrate strength in visual-spatial and geometric reasoning. Others are creative problem solvers or excel when given tasks that require the use of conceptual understanding over rote repetition. These strengths matter to getting a clear picture of who our students are.

Conversely, not all students who struggle with math have dyscalculia. Any number of things can impact a student's math learning. For example, a student with dyslexia may demonstrate difficulty interpreting word problems, while a student with ADHD may forget steps in a task. A student with autism might struggle to move flexibly between strategies, while geometry and place value may be a challenge for students with spatial difficulties.

When we move past misconceptions, we strengthen our ability to identify and respond to dyscalculia with clarity and confidence, ensuring that our teaching is grounded in understanding rather than myth.

A Clearer Lens For Complex Struggles

As educators, we're used to seeing students who struggle with math. But without a common language, it can be easy to misunderstand those struggles, attributing them to effort, maturity, or gaps in instruction alone. This chapter offers a more precise lens, drawing on research and guidance from the DSM-5 and the SASC to help us recognize and respond to students with potential dyscalculia.

Two **key components** emerged: **numerical magnitude processing** – including symbolic and non-symbolic thinking, estimation, counting, and sequencing- and mathematical vocabulary – encompassing both technical terms and everyday language with mathematical meaning. These aren't abstract concepts for clinicians. They're patterns we see every day in our classrooms: the student who thinks 0.9 is smaller than 0.45, or the one who can't tell whether to add or subtract in a word problem.
While diagnosis isn't our role, our observations matter. We are often the first ones to notice when something deeper is going on. With the right language and understanding, we can shift from guessing to knowing; offering students not just support, but clarity, dignity, and access to learning that feels possible.

Reflection Questions for Teams and Book Study Groups

These questions can be used to guide discussion or individual reflection.

1. What did you learn about dyscalculia that surprised you or challenged your assumptions?
2. How do the SASC's two essential components: numerical magnitude processing and mathematical vocabulary, connect to what you have seen in students before?
3. Which students come to mind as you read about symbolic/non-symbolic magnitude processing, estimation, counting, or sequencing? How might this new lens change your instructional approach?
4. How might the language of math, both formal terms and everyday words, show up as a barrier for students in your classroom? What could you do to make it more accessible?
5. How can a more precise understanding of dyscalculia help your school avoid both overidentifying and overlooking students with math difficulties?

References

American Psychiatric Association. (2013). Diagnostic and statistical manual of mental disorders (5th ed.). https://doi.org/10.1176/appi.books.9780890425596

Pagliaro-Newman, D. (2024, April 25). *Shining A Light On Dyscalculia*.Edutopia. https://www.edutopia.org/article/teaching-students-dyscalculia

SpLD Assessment Standards Committee. (2025, March 24). *Guidance on the assessment of mathematics difficulties and dyscalculia.* SASC

Chapter 2 Dyscalculia In The Classroom

Leo sits at his desk in his second-grade classroom, a pile of linking cubes gathered in front of him, while his teacher observes.

"One, two, three…" he recites, brow furrowed, dragging each cube slowly across the table. "…Four, five—no, wait!"

He pauses, looks back at the cubes he's already moved, and starts over for the third time, eventually arriving at the total.

"Thirteen," he declares.

Gripping his pencil tightly, Leo records his answer: 31.

The reversal of terms like *thirty* and *thirteen* is a common one for him. But still, his teacher sighs with disappointment—he seemed to have it just yesterday.

"Why is Leo's language, writing, and understanding of place value still so tangled?" she wonders. "And by the time he finally writes his work down, it's like his thoughts get stuck in a traffic jam."

Across the city, in a fifth grade classroom sits Sophia, fingers pressed to the desk. "Ten, eleven, twelve," she taps as she whispers, arriving at the product of 3 and 4. She doesn't have to look up to know that the others have already completed the warm up and moved on to the assignment of the day.

One problem reads: *Four people share 1/2 of a bag of potato chips. How much does each person get?*

"Crumbs, probably," she thinks.

With a groan, Sofia reads the problem again, her eyes scanning the numbers but not quite landing on their meaning. She knows how to divide. She's seen fractions before. But this time they're combined and inside a word problem. It's like trying to follow a map written in a language she almost understands. She doesn't know where to start.

As her classmates are finishing up, her paper is still blank. It's not that she's not trying, it's because the steps she knows don't seem to fit. She can't picture what is happening in the problem, and the connection between the numbers, the story, and the operation remains just out of reach.

Take A Moment To Notice

Their names are fictional, but the stories of Leo and Sofia are rooted in the shared experiences of many students across years and grade levels with whom I have worked, much like the ones sitting at your small group table, staring at a blank worksheet, or trying (and trying again) to get it right.

- What parts of their struggle feel familiar to you?
- Do you recognize any of your students in them?
- What strengths do you see, even in the midst of difficulty?
- What might you try to better understand what is going on beneath the surface?

This next section introduces a lens that can bring clarity to times like these. But before we get there, let's pause. Make a mental note, or a written one, of a student you are thinking about.

Understanding Subtypes: A Practical Lens For Teachers

There is no single, universally agreed upon system for classifying dyscalculia into subtypes. Researchers have proposed different ways of grouping characteristics, each grounded in their own theoretical frameworks and goals. Some focus on underlying cognitive processes, others on brainbased differences, and still others on patterns of errors or academic performance.

But in the classroom, we have a wide range of learners to teach and a mandated curriculum to follow. Curriculums move quickly, there never seems to be enough time to cover it all, and every lesson reveals another misconception to address. At times, the list of student struggles becomes overwhelming and deciding where to start can be immobilizing.

Leo is a perfect example of the overwhelm. He tries so hard, carefully lining up the linking cubes, counting out loud, restarting when he gets confused. Yet after all that effort he writes 31 instead of 13. His teacher knows he is capable, but his language, writing, and place value understanding always seem to get tangled. She wants help, but where does she begin?

Then there's Sofia. She's older, quieter. From the outside it may look like she's not trying, but that couldn't be further from the truth. She reads the word problem, then reads it again. She knows how to divide. She's worked with fractions. But now they're combined in a way that doesn't quite make sense. She draws a blank. The math she does know doesn't seem to apply here. Her page stays empty, even though she has been working the whole time.

It's understandable to look at students like Leo and Sofia and feel stuck. Their struggles seem so different, yet equally hard to untangle. Are Leo's challenges about language? Conceptual understanding? Or is the act of writing itself what is slowing him down? And what about Sofia? Are her difficulties rooted in fractions? Word problems? Processing speed? Each student error can feel like a separate issue, leaving us uncertain where to begin.

That's why we need to look for patterns, not to diagnose, but to make sense of what we are seeing. We need to shift our thinking from "This student is struggling with *this, this,* and *this,*" to "This student is struggling with this *type of math thinking*." That shift helps us plan more intentionally and teach more effectively.

Researcher Giannis Karaginannakis and colleagues (2014) have taken these patterns and proposed four subtypes of dyscalculia that closely align with common domains in elementary math, making it easy to connect to the typical scope and sequence of a school year. These subtypes reflect the types of thinking that often breaks down for students with dyscalculia and mirror how curriculums tend to be structured. That makes them a practical starting point for deciding where to focus our instructional energy.

Other researchers have proposed their own subtype models as well, reflecting different ways of understanding how mathematical thinking can break down. Some focus on underlying cognitive processes, others on developmental stage. None of these subtypes are universally agreed upon, but for our purposes, using one model gives us a common language. It allows us to organize observations around recognizable patterns, not labels.

Table 2.1: Classification Of Mathematical Learning Difficulties Subtypes

Subtypes and Math Difficulties	
Core Number	**Memory**
• Subitizing • Counting • Estimating • Meaning of place value (including fractions, decimals, and percents) • Understanding the meaning of operation symbols (+, -, x, ÷) Placing numbers on number lines • Changing rational numbers (whole numbers, fractions, decimals and percents) from one form to another • (numbers, pictures, and words)	• Remembering number facts • Remembering vocabulary (numerator, equivalent…) • Making sense of spoken directions and turning them into steps numbers, or written work • Performing mental math accurately • Remembering and carrying out rules and procedures • Keeping track of steps when problem solving
Reasoning	**Visual-Spatial**
• Understanding mathematical ideas and how they connect to one another • Understanding multiple steps in procedures and/or algorithms Grasping basic logic (if…then…) • Decision making when solving problems	• Making sense of how shapes and numbers are organized on the page • (place value, exponents, parts of a shape…) • Placing numbers on a number line • Being able to tell math numbers and symbols apart • Solving work on paper where place value or organization is important • (carrying, borrowing…) • Filtering out unimportant information • Making sense of shapes, especially rotations • Interpreting graphs and tables

Note. Adapted from *"Mathematical learning difficulties subtypes classification,"* by G. Karagiannakis, A. BaccagliniFrank, & Y. Papadatos, 2014, *Frontiers in Human Neuroscience, 8,* 57 (https://doi.org/10.3389/fnhum.2014.00057). CC BY license.

Taking A Closer Look

Let's examine each subtype with greater depth, not to rigidly categorize our students, but to identify patterns across tasks and contexts. Most students will show overlapping difficulties across more than one subtype. Still, recognizing where specific types of thinking consistently break down can offer valuable insight and serve as a starting point for targeted instructional planning.

We begin with core number, a domain that underlies many early math experiences and often reveals some of the earliest signs of dyscalculia.

<u>Core Number</u>

What It Is

Core number refers to the foundational skills students use to understand and work with quantity. This includes subitizing, counting accurately, comparing numbers, estimating, understanding place value, interpreting math symbols, and switching between different representations (like digits, words, or visuals).

These are the concepts that help students build number sense, a flexible intuitive understanding of how numbers work. This "innate number system" has been called 'the number module" by cognitive neuroscientist Brian Butterworth (2019). He defines dyscalculia as "a deficit in the number module." In other words, when students struggle with core number, it often means that this internal sense of quantity, the ability to "feel" and work with numbers intuitively, isn't as reliable for them. They may not trust numbers, notice patterns, or grasp the relationships that make math make sense.

Without this strong, internal sense of number, students often rely on memorization or procedures without truly understanding the quantities behind them, creating a fragile foundation that can lead to confusion as math becomes more complex.

Why It Matters

When students struggle with core number, nearly every area of math becomes harder. Difficulties may not be obvious at first, especially for students who are good at memorizing and mimicking. But because their underlying number sense is fragile, as content becomes complex, cracks begin to show, especially with mental math, multistep problems, and confidence. Core number difficulties will persist if not directly addressed. The good news is that this subtype responds well to visual, concrete, and language rich supports, approaches already familiar to elementary educators.

Classroom Clues

Students with core number difficulties might:

- Struggle to count accurately, or lose track while counting objects.

- Rely heavily on one-to-one counting strategies, even when inefficient.

- Confuse similar looking numbers (13 vs 31) or misread multidigit numbers.

- Misunderstand or misuse basic symbols.

- Place numbers inaccurately on a number line.

- Struggle to compare two numbers meaningfully ("Which is more, 46 or 64?").

- Have difficulty showing numbers in more than one way (pictures, digits, words).

In Leo's case, the reversal of 13 and 31, the repeated restarting while counting, and the slow transition from manipulatives to paper, all point to a fragile number sense, particularly around place value and representation. He works hard, but the basic structure of number remains elusive. In upper elementary and middle school, students with core difficulties show additional challenges. They may:

- Misunderstand rational numbers: whole numbers, fractions, decimals and percents.

- Represent rational numbers in only one way (pictures, words, or numbers) making word problems very challenging.

- Misunderstand how operations (addition, subtraction, multiplication, division) effect rational numbers.

- Misunderstand the equal sign as giving an answer, as opposed to a symbol that shows an equivalent relationship between terms.

Sofia's struggles reflect the kinds of challenges that arise for students with core number difficulties in upper elementary. When faced with the problem of sharing half a bag of chips among four people, she cannot connect the story to the numbers or represent the fractions in more than one way. Even with multiplication facts like 3×4, she relies on finger counting, showing that relationships and equivalence remain uncertain.

Try This

To support students with core number challenges:

- Use manipulatives like linking cubes on ten frames, base ten blocks, or fraction strips to build meaning through touch and movement.

- Incorporate subitizing practice using dot cards and ten frames. Flash briefly, then ask to describe what they saw.

- Emphasize number structure by breaking multi-digit numbers. Ask questions such as: How many hundreds? Tens? Ones?

- Use Number Strings, a series of related problems designed to build fluency with a specific strategy (making ten, compensation, etc.). Number Strings offer explicit, guided practice and are especially helpful when students are still developing new strategies.

- Explore Number Talks. Once a few strategies are secure these open-ended problems invite students to choose and explain strategies, strengthening their flexibility and mathematical discourse.

- Connect representations by having student match digits to visual models, write numbers in words, or sketch what a quantity looks like.

- Encourage estimation in daily routines by asking simple questions such as "Do you think there are more than 50 cubes in here or fewer?"

- Reteach math symbols explicitly with examples and non-examples. Reinforce them with visuals or movement.

Number Strings and Number Talks: When and Why

Two powerful routines, both grounded in meaningful discourse, but best used at different stages of the learning process.

Start with Number Strings

Created by Cathy Fosnot, Number Strings are carefully sequenced sets of related math problems designed to model a specific mental math strategy. The teacher presents each problem one at a time, inviting students to share how they got their answer. Each Number String is designed to focus on one strategy, allowing students to work towards mastery.

Number Strings aren't just about getting answers, they highlight structure, strategy, and efficiency. For students with dyscalculia, this structure is especially important. Number Strings reduce cognitive load by narrowing focus to a single idea and offering repeated, supported practice.

Use Number Strings when students need explicit modeling and guided practice with a strategy they haven't yet internalized. This is especially helpful for those who struggle with flexible number thinking. Two powerful routines, both grounded in meaningful discourse, but best used at different stages of the learning process.

Start with Number Strings

Created by Cathy Fosnot, Number Strings are carefully sequenced sets of related math problems designed to model a specific mental math strategy. The teacher presents each problem one at a time, inviting students to share how they got their answer. Each Number String is designed to focus on one strategy, allowing students to work towards mastery.

Stretch with Number Talks

Sherry Parrish's Number Talks take that foundation and open it up. Similarly to Number Strings, Number Talks are carefully designed sets of related problems. However, in a Number Talk, students choose their own strategies, compare approaches, and explain their reasoning. Number Talks are designed in a way that allows for the use of multiple strategies, as opposed to Number Strings that focus on one at a time. Number Talks are designed to allow students to find their own meaning and find relationships. While this can be challenging for students with dyscalculia, it can also be empowering, offering them opportunities to hear varied strategies, make connections, and build confidence, as they contribute in ways that make sense to them.

Use Number Talks when students are ready to explore and reflect on different approaches, and when the goal is to deepen reasoning, not increase speed.

Number Strings put tools in the toolbox. Number Talks use them. For students with dyscalculia, both steps matter. Number Strings aren't just about getting answers, they highlight structure, strategy, and efficiency. For students with dyscalculia, this structure is especially important. Number Strings reduce cognitive load by narrowing focus to a single idea and offering repeated, supported practice.

Use Number Strings when students need explicit modeling and guided practice with a strategy they haven't yet internalized. This is especially helpful for those who struggle with flexible number thinking.

<u>Memory</u>

What It Is

Students with memory-based difficulties often know more than they can show. They might forget steps in the middle of solving a problem, lose track of directions, or pause to recall a word like denominator even when they understand what it means. Facts like 6x7 don't always come easily, and strategies can slip away just when they're needed. These challenges are related to both working memory, holding on to information in the moment, and long-term memory, retrieving information learned in the past.

It's important to remember that these memory challenges aren't about effort or attitude. Students with dyscalculia often work hard to learn something, and then find it's gone the next day. They may need more time, more exposure, or different pathways to hold on to information. It's not that they aren't trying. It's that their brain handles the storage and retrieval of math in a different way.

Why It Matters

Memory is the glue that holds math learning together. It connects the steps in a procedure, helps students recall the meaning of symbols, and allows them to apply strategies across contexts. When memory breaks down, students may work inefficiently, forget critical information, or struggle to complete problems they otherwise know how to solve.

More than that, inconsistent memory retrieval often leads to confusion and self-doubt. A student who feels like they are constantly forgetting or falling behind may stop participating or attempting challenging tasks altogether.

Leo's mix-up between thirty and thirteen, especially after seemingly mastering it, is one small example of this. He knows the idea, but the verbal label slips. It's not a lack of effort. It's a difficulty retrieving and recalling what he knows. The same is true for Sofia, who uses her fingers to solve 3x4, not because she doesn't understand multiplication, but because the facts just won't stick. Both students know more than they can easily show, especially when memory gets in the way.

Classroom Clues

Students with memory difficulties might:

- Rely on slow or inefficient strategies (like finger counting or drawing repeated pictures).
- Lose their place in multistep problems.
- Forget procedures soon after learning them.
- Frequently ask for reminders about directions or definitions.
- Confuse math vocabulary or switch similar sounding terms.
- Seem inconsistent. (Confident one day, confused the next).
- Become anxious or resistant when expected to recall facts quickly.
- Hesitate to begin working, especially when unsure where to start.

Try This

To support students with memory challenges:

- Use visual anchors like charts, tables, and word walls.
- Limit memory load by breaking tasks down into smaller steps.
- Teach concepts explicitly, rather than relying on the memorization of procedures.
- Incorporate retrieval practice in low pressure ways such as games and warm ups.
- Praise strategy, not just speed or accuracy.

<u>Reasoning</u>

What It Is

The reasoning subtype refers to difficulties understanding and applying connections between mathematical concepts, making it difficult to see how it all fits and works together. This includes choosing an operation when solving a word problem, getting lost in the steps of a multistep process, or difficulty explaining how operations work. Students with reasoning-based difficulties tend to need additional support learning to problem solve flexibly and generalize patterns.

For students with dyscalculia, the reasoning subtype might show up as difficulty identifying the relationship between numbers or relying heavily on memorized steps. They may also struggle to explain their thinking or recognize when an answer does not make sense.

Why It Matters

Mathematics is more than just computation. It's a discipline of reasoning. When students can't understand the "why" behind a concept, they are less likely to retain it, recognize their errors, and make connections to other concepts. This impacts not just accuracy, but agency as well. This leaves students feeling less confident and less willing to take risks.

Students like Sofia are often quietly stuck here. It appears as if she understands the concepts. She can find a quotient. She can represent fractions. But once it is layered together within a context, or shown to her in an unfamiliar format, she is unable to make sense of it. Her struggle isn't procedural, it's conceptual.

Classroom Clues

Students with difficulty reasoning might:

- Memorize steps without understanding why they work.
- Struggle to decide which operation to use in word problems.
- Miss logical relationships. (E.g., If 7+3=10, then 7+4 is one more.)

- Struggle to explain or justify their reasoning.
- Apply a strategy in the wrong context.
- Find multi-step problems overwhelming or confusing.
- Give up when a problem is unfamiliar.

Try This

To support students with reasoning-based difficulties:

- Model and narrate your thinking when solving problems aloud.
- Use drawings, lists, or manipulatives to make patterns and relationships visible.
- Ask questions like, "What do you notice?"
- Show examples of student work and ask, "Why might this person have solved it this way?"
- Explicitly connect new concepts to previous learning.

Sofia At The Small Group Table

Sofia sits with her math group, eyes on the whiteboard.

Teacher: Okay, here's the problem. A recipe uses ¾ cup of sugar. You want to make half the recipe. How much sugar do you need?
Sofia furrows her brow.
Sofia: So it's half of ¾?
Teacher: Exactly. What do you think that might be?
Sofia: (quietly) Um…maybe 2?
Teacher: Let's think through that. Is 2 more or less than ¾?
Sofia: More?
Teacher: But you're taking half of ¾. So shouldn't the answer be smaller?
Sofia: (nodding slowly) Yeah. I guess… I just… Like, is it ¾ divided by 2, or do I do something with the denominator?
She glances at her classmate's paper and then back at her own.

Take A Moment To Notice

Sofia remembers learning something like this before, but she can't reason what "half of ¾" actually means. When faced with a twist on a familiar topic, she is unsure how to proceed. This is the heart of the reasoning subtype: knowing the pieces, but not being able to connect them.

What do you notice about Sofia's thinking?

- What does she understand?

- Where does she get stuck?

- What clues does she give about her thinking?

If you were the teacher in this moment, how might you respond?
Would you:

- Use a visual or model?

- Ask a new question?

- Return to a familiar context?

- Invite peer thinking or group discussion?

One Way The Conversation Might Continue

Sofia holds her pencil above her paper, unsure where to begin.

Teacher: You're right to pause here. This part's tricky. What do we know from the problem?
Sofia: There's 3/4.
Teacher: 3/4 of what?
Sofia: (looking back her paper) 3/4 of a cup of sugar.
Teacher: (drawing a rectangle partitioned into four pieces) Right. That's what we know from the problem. Now, let's picture it. What would that look like?
Sofia: (coloring three parts of the rectangle) That's 3/4.
Teacher: Yes! Now remember how you said the question is asking for half of 3/4? What would that look like on the picture?
Sofia leans forward and draws a line through all four pieces.
Teacher: Now what do you see?
Sofia: It's 6/8.
Teacher: Yes. The picture shows 6/8. The recipe only needs half.
Sofia: (pointing to the picture) You only need this part. It's 3/8.
Teacher: Yes! Half of 3/4 is 3/8. You used a picture to make sense of the problem.

Why This Matters
Instead of correcting Sofia, or giving her a rule, the teacher broke the problem down into more accessible parts, while providing a visual context by drawing a picture. This allowed Sofia to arrive at the answer, proving that she can reason through a challenge with the right support.

<u>Visual – Spatial</u>

What It Is
The visual-spatial subtype can be seen when students have trouble understanding how numbers, symbols, or shapes are

organized on the page. This might mean struggling to line up digits in multi-digit multiplication, misreading place value, or losing track of where to plot a point on a graph. It can also show up in geometry tasks when students are asked to make sense of how a shape turns or flips, or in data work, like creating tables or charts. For these students, it's not about being messy or careless, It's about how their brain processes position and direction. What looks like a simple mistake can be a deeper challenge with spatial organization.

Why It Matters

Whether it's where digits are placed or how equations are arranged, structure in math matters. When students misinterpret this structure, it affects accuracy, comprehension, and confidence. These students may have strong verbal or reasoning skills, but still underperform because of issues with spacing, orientation, or symbols confusion.

A student like Leo might reverse digits, writing 31 instead of 13, misalign columns when adding or subtracting, or confuse < and >. What may seem like a careless mistake could be the result of visual-spatial processing challenges.

Classroom Clues

A student with visual-spatial difficulties might:

- Reverse digits or symbols (e.g. 6 and 9, < and >).
- Struggle with aligning numbers for vertical calculations.
- Misplace digits in place value (e.g. writing 14 or 1004 instead of 104).
- Appear disorganized in written work.
- Struggle to read or create graphs, charts, or tables.
- Have difficulty copying from the board.
- Struggle with rotation or transformation tasks in geometry.

Try This

To support students with visual-spatial difficulties:

- Provide graph paper or templates to help align digits in computation.
- Color code place value columns for clarity.

- Model how to read and interpret graphs and visual models.
- Use hands-on manipulatives to support abstract ideas.
- Teach students to circle important information or symbols before starting.
- Provide lots of practice copying from different visual sources.

Putting It All Together: Using Subtypes to Understand Your Students

Each subtype offers a unique lens for understanding math difficulties, but their real power comes from using them together; not as labels but as a way to notice patterns across tasks and time.

Students rarely struggle in just one area. More often we see overlapping difficulties: a student who misplaces digits (visual-spatial) and forgets procedures (memory), or another who has difficulty understanding decimal notation (core number) and freezes when asked to solve word problems (reasoning). By looking for patterns instead of isolated mistakes we begin to build a student profile that reflects the underlying types of thinking that are breaking down. This shift is important. Instead of addressing every individual error, we start to see why those errors are happening. And that gives us a clearer path forward.

Building A Student Profile: When and Where To Begin

This process is not for every student, nor should it be. Creating a student profile is most helpful when a student continues to struggle despite the typical supports we put in place. These are the students who leave us asking, "What am I missing?" The students who works so hard, but makes minimal progress, or the one who seems to improve for a short time but then slips back. When traditional interventions and reteaching does not lead to meaningful progress, it is a signal that we need to look more closely, not to label the student but to understand how they are making sense of mathematics. Creating a profile helps us *pause and see the thinking patterns* underneath.

Chances are you have already been thinking of a few students. Choose one of them and think about what you have observed. Don't worry about getting it perfect. This isn't a diagnosis, remember? This is just a starting point: a tool not for a solution, but for a direction.

Ask yourself:

- What kinds of tasks tend to break down for this student?
- What specific errors keep repeating, even after reteaching?
- Do these errors seem related to
  - Core Number (quantity, place value, basic operations)?
  - Memory (fact and procedure retrieval)?
  - Reasoning (connecting ideas and choosing strategies)?
  - Visual-Spatial (misalignment, symbol confusion, graphs)?

Don't worry about fitting everything into neat categories. You may even notice that you don't have any errors for some of the subtypes. The goal is simply to look for reoccurring types of difficulty that can inform our instruction.

A Planning Tool, Not A Label

This process isn't about identifying "what kind of student" you have. It's about identifying what kind of support the student may need. When you recognize that a child's difficulty with subtraction is not just about regrouping - it's about remembering the steps and keeping the digits aligned - you can make strategic decisions about how to intervene. We can use these four subtypes to organize our observations and bring clarity to the messy real-world work of teaching students who struggle with math.

Take Leo, for example. At first glance his challenges might seem scattered: counting errors, reversed digits, incorrect answers that seem far off from the problem. But when we step back and map out what we are seeing through the lens of the four subtypes, patterns begin to emerge. Let's use a checklist to move from a collection of disconnected mistakes to a clear picture of the types of thinking that challenges Leo.

Student Error Pattern Checklist (Sample For Leo)

Student Name: _____ _LEO_ _____ Date: _____ 11/16 _____
Grade/Class: _____ SECOND GRADE _____

Directions: Check all that apply as you review student work, observe during instruction, or reflect on student responses. Look for clusters. They may point to where breakdowns are happening most consistently.

Core Number

- ☒ Counts slowly or loses track when counting.
- ☐ Struggles with subitizing (recognizing small amounts without counting).
- ☒ Confuses similar numbers (e.g. 13 and 31, 42 and 24, 101 and 110).
- ☐ Difficulty with hierarchical inclusion (understanding that a number is made up of smaller numbers).
- ☐ Struggles to count backward or start at a number that is not zero or one.
- ☒ Struggles to represent quantity in different ways (numbers, pictures, and/or words).
- ☐ Difficulty rounding or estimating.
- ☐ Difficulty with place value understanding.

Upper Elementary & Middle School

- ☐ Struggles to demonstrate a conceptual understanding of rational numbers (whole numbers, fractions, and decimals).
- ☐ Has limited ways to represent rational numbers (whole numbers, fractions, and decimals).
- ☐ Unable to use multiple representations of rational numbers (whole numbers, fractions, and decimals) to solve problems.
- ☐ Difficulty understanding how operations (addition, subtraction, multiplication, division) effect rational numbers (whole numbers, fractions, and decimals).
- ☐ Understands the equal sign as giving and answer as opposed to a symbol that shows an equivalent relationship between terms.

____

____

Memory
- ☐   Uses fingers or manipulatives for simple calculations.

- ✗   Difficulty performing mental math.
- ✗   Forgets previously learned procedures.
- ☐   Difficulty recalling learning number facts.
- ☐   Difficulty remembering mathematical vocabulary.
- ☐   Struggles to move from concrete to abstract representations.
- ✗   Seems to get lost in the steps of a procedure.

____

____

Reasoning
- ☐   Difficulty explaining thinking.
- ☐   Struggles to make connections between concepts.
- ✗   Reliance on rote learning.
- ☐   Applies the same strategy regardless of problem type.

When we look at Leo's checklist, we begin to see how multiple challenges are clustered within the Core Number and Memory subtypes. He struggles with counting and representing quantities, all foundational aspects of number sense. Add that to his reliance on manipulatives for simple problems, and difficulty keeping track of steps, and we start to see that his surface level errors reflect deeper challenges with how he understands and holds on to numerical information. This does not give us a diagnosis. This gives us a direction. Rather than trying to fix every mistake at once, we can focus our energy on strengthening his number sense, building fluency with small quantities, and using memory supports that help him hold on to procedures over time.

Now let's explore what patterns we can find in Sofia's errors. Sofia's difficulties don't always look like Leo's. She is quiet, which makes it easy to overlook her in a busy classroom. She rarely asks for help, even when she is stuck, and will often wait until others have answered before attempting to solve a problem herself. On the surface, her work looks neat and careful, but it's filled with inconsistencies. One day she can solve a multi-step problem with minimal support, the next she freezes on a basic fact. These ups and downs can make it hard to pin down where the breakdown is really happening. But when we zoom out and use the same checklist we start to notice patterns. Student Error Pattern Checklist (Sample For Sofia)

Name: _____ *Sofia* _____ Date: _____ 10/21 _____
Grade/Class: _____ *5th grade* _____

Directions: Check all that apply as you review student work, observe during instruction, or reflect on student responses. Look for clusters. They may point to where breakdowns are happening most consistently.

Core Number
- ❑ Counts slowly or looses track when counting
- ❑ Struggles with subtilizing (recognizing small amounts without counting)
- ❑ Confuses similar numbers (e.g. 13 and 31, 42 and 24, 101 and 110)
- ❑ Difficulty with hierarchical inclusion (understanding that a number is made up of smaller numbers)
- ❑ Struggles to count backward or start at a number that is not zero or one.
- ❑ Struggles to represent quantity in different ways (numbers, pictures, and/or words)
- ❑ Difficulty rounding or estimating
- ❑ Difficulty with place value understanding

Upper Elementary & Middle School
- ❑ Struggles to demonstrate a conceptual understanding of rational numbers (whole numbers, fractions, and decimals)
- ✗ Has limited ways to represent rational numbers (whole numbers, fractions, and decimals).
- ✗ Unable to use multiple representations of rational numbers (whole numbers, fractions, and decimals) to solve problems.
- ✗ Difficulty understanding how operations (addition, subtraction, multiplication, division) effect rational numbers (whole numbers, fractions, and decimals).
- ❑ Understands the equal sign as giving and answer as opposed to a symbol that shows an equivalent relationship between terms.

____

____

Memory
- ✗ Uses fingers or manipulatives for simple calculations
- ✗ Difficulty performing mental math
- ✗ Forgets previously learned procedures
- ✗ Difficulty recalling learning number facts

> ❏   Difficulty remembering mathematical vocabulary
> ❏   Struggles to move from concrete to abstract representations
> ✗   Seems to get lost in the steps of a procedure
>
> ____
>
> ____

When we look at Sofia's checklist, a more complex profile begins to emerge. While she shows clear difficulties with Core Number, her challenges also extend to Memory and Reasoning. Her surface-level work may appear tidy, but the checklist reveals deeper inconsistencies in how she stores and applies mathematical information. She relies heavily on fingers and manipulatives for simple calculations, forgets previously taught procedures, and struggles with mental math and fact recall—clear indicators of memory-related difficulties. These are compounded by reasoning challenges: she has difficulty explaining her thinking, connecting concepts, and making decisions during problem solving. Her reliance on rote steps often masks a lack of conceptual understanding. Unlike Leo, Sofia shows fewer visual-spatial or core number difficulties; her errors are less about how numbers are perceived and more about how they are retained, accessed, and used flexibly. This profile helps us shift away from reteaching procedures in isolation. Instead, we can focus on helping Sofia build connections between ideas, use strategy-based fluency routines, and develop memory supports like visuals, verbal cues, and guided practice that strengthens both recall and reasoning.   When students struggle with math, the challenges can feel scattered and hard to define. But by looking through the lens of subtypes: Core Number, Memory, Reasoning, and Visual-Spatial, we begin to see patterns emerge. These patterns offer us a practical entry point for understanding our students and making intentional choices about how we teach.

The Student Error Pattern Checklist is one way to organize our observations and clarify where breakdowns are happening most consistently. In later chapters, we will return to this information as we explore how to plan small group lessons, adapt core instruction, and select the right scaffolds for support. But before we do, we'll take one more step in building our student profiles. In the next chapter, we'll focus on something just as important as understanding a student's challenges: recognizing and documenting their strengths.

Reflection Activity for Teams and Book Study Groups

As a group, select a student from your own setting. Use the Student Error Pattern Checklist in Appendix A to begin a profile together.

- What patterns do you notice across the four subtypes?
- Which subtypes seem most connected to this student's math struggles?
- How might identifying these patterns shape your next steps for instruction or intervention?

References

Butterworth, B. (2019). *Dyscalculia: From science to education*. Routledge.

DuPaul, G. J., Gormley, M. J., & Laracy, S. D. (2013). Comorbidity of LD and ADHD: Implications of DSM-5 for assessment and treatment. *Journal of Learning Disabilities, 46*(1), 43–51. https://doi.org/10.1177/0022219412464351

Fosnot, C. T., & Uittenbogaard, W. (2007a). *Minilessons for early addition and subtraction*. Heinemann.

Karagiannakis, G., Baccaglini-Frank, A., & Papadatos, Y. (2014). Mathematical learning difficulties subtypes classification. *Frontiers in Human Neuroscience, 8*, 57. https://doi.org/10.3389/fnhum.2014.00057

Parrish, S. (2014). *Number Talks Common Core Edition, Grades K-5: Helping Children Build Mental Math and Computation Strategies*. Math Solutions.

Chapter 3 Identifying Strengths

When we lead with the belief that every student is a capable mathematician, we start to look for strengths as carefully as we look for errors. This shift is especially important for students with dyscalculia and other learning differences. Too often, disability is viewed as something broken within the student. But in reality, it often reflects a mismatch between a learner's strengths and their educational environment.

In her book, *Rethinking Disability and Mathematics*, Rachel Lambert writes, "Disability seems to not only be part of an individual student but also what is valued in their math class, demonstrating how disability can be constructed socially in mathematics classrooms." In other words, what we notice, prioritize, and celebrate in our classrooms plays a powerful role not only in shaping how students are seen, but also how they see themselves.

When we pay attention to student strengths, we shift the narrative. We begin to recognize that mathematical brilliance doesn't always show up in predictable ways. Some students may visualize relationships that others can't see. Some may reason with flexibility and insight, even if they struggle with fluency. Some may persist long after others have given up. These strengths matter and they must inform the way we plan, teach, and support our students. The next step in building a student profile is an important one. By identifying and documenting student strengths, we identify where instruction begins.

Expanding Our View of Mathematical Brilliance

What does it mean to be mathematically proficient? Too often our definition is narrow; focused on speed and accuracy. But math is more than just right answers, it involves reasoning, representation, strategic decision-making, and perseverance.

The Strands of Mathematical Proficiency, developed by the National Research Council in *Adding It Up* (2001), offers a more inclusive definition. Five interwoven strands - Conceptual

Understanding, Procedural Fluency, Strategic Competence, Adaptive Reasoning, and Productive Disposition - provide a research-based lens for recognizing and supporting students. They give us language to describe what students can do so we can ensure that those strengths are reflected in our planning. Just as identifying patterns of difficulty can bring clarity to student needs, the Strands of Mathematical Proficiency help us recognize patterns of strength that might otherwise go unnoticed. They shift our focus from isolated skills to broader ways of thinking, helping us see students as capable, complex mathematical thinkers even when their abilities don't align neatly with grade level expectations.

Figure 3.1: The Strands of Mathematical Proficiency

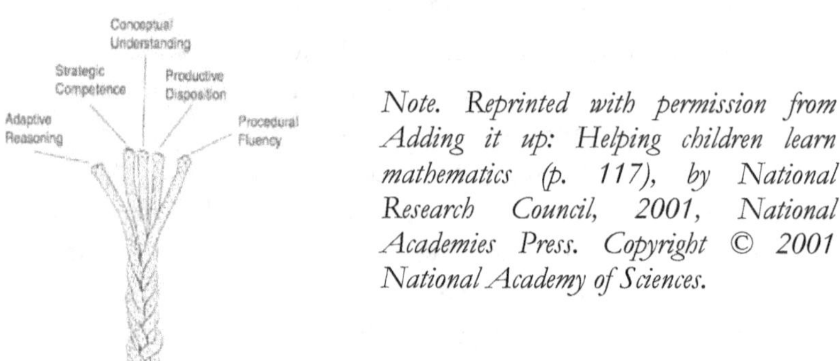

Note. Reprinted with permission from Adding it up: Helping children learn mathematics (p. 117), by National Research Council, 2001, National Academies Press. Copyright © 2001 National Academy of Sciences.

In the sections that follow, we will explore each strand, consider how it might appear in student work and behavior, and discuss how these strands can support our understanding of students with dyscalculia. By focusing on what students are already doing well, we can build bridges to the areas where they need more support.

Conceptual Understanding

Conceptual understanding doesn't just mean knowing what a math concept is. It means understanding how ideas relate to each other and why procedures work. According to *Adding It Up* (NRC, 2001), conceptual understanding involves having a deep and flexible knowledge base. Students with conceptual understanding can see how ideas connect and understand how one concept builds on another. They recognize patterns, make generalizations, and explain their thinking. Because their understanding is rooted in meaning rather than memorization, they are more likely to adapt when faced with an unfamiliar problem.

For students with dyscalculia, conceptual understanding is especially important. It can serve as a strength that helps compensate for difficulties with memory or visual-spatial processing. For example, a student may not recall their multiplication facts, but they do understand that one meaning of multiplication is repeated addition. They may use that knowledge to build equal groups with counters, skip count, or draw an array. Even if the final answer takes longer to reach, their conceptual understanding is sound and that becomes the bridge to more efficient strategies over time.

Teachers are trained to look for errors so we can help students improve for good reason. But when that becomes our only lens, it becomes very easy to miss what is working. By intentionally looking for signs of conceptual understanding, we give students with dyscalculia the opportunity to engage meaningfully with mathematics, even when fluency or organization is a challenge.

Practicing a Strengths-Based Mindset

A strengths-based mindset asks us to pause, look more closely, and shift the narrative. Try using questions such as:

- What parts of the task did the student approach with confidence or creativity?

- What kinds of tools, models, or strategies does the student reach for independently?

- What mathematical ideas does the student use accurately in conversation, even if it's not reflected in their written work?

These kinds of questions help us move beyond a list of deficits and begin to recognize the full range of what a student can do.
Instead of saying, "He can't memorize his facts," we might say, "He understands multiplication as repeated addition and can represent it with models."
Could they have eaten the same amount? *How do you know?*

> Instead of saying, "She mixes up the steps every time," we might say, "She's trying out new strategies and is beginning to notice when they don't work."
>
> This kind of thinking takes practice, but it helps us see students more clearly, plan more effectively, and build stronger math identities.

Procedural Fluency

When someone says their child is "good at math", they often follow up with, "They already know their math facts." Procedural fluency is often mistaken for speed or memorization and is sometimes treated as the ultimate marker of math success. But according to *Adding It Up* (NRC, 2001), true procedural fluency is much more than that. It refers to the ability to carry out procedures accurately, flexibly, and efficiently, including basic computation, mental math, and algorithms. Students who are procedurally fluent can estimate the reasonableness of their answer, select appropriate strategies, and adjust methods when needed.

Mathematician and professor Anna Stokke calls math, "relentlessly hierarchical" and believes that procedural fluency "reduces cognitive load, which frees up space for your brain to concentrate on more complex problems in say algebra or geometry" (Stokke, 2024). Yet, despite its importance, some educators worry that a focus on fluency comes *at the expense* of deeper understanding.

The reality is that procedural fluency and conceptual understanding are not in conflict. They are deeply intertwined. When students understand why a procedure works, they are more likely to remember it, apply it accurately, and recognize when it breaks down. Likewise, fluency practice can help students uncover patterns and relationships that deepen their conceptual understanding.

Fluency does not mean rote memorization without meaning. It means solving problems accurately, efficiently, and flexibly, knowing when and how to use a strategy. Some students with dyscalculia may, in fact, have strong memories and use fact recall to their advantage. For them, memorization can be a powerful strength. However, it is more common for students with dyscalculia to struggle with fact retrieval, even after repeated exposure and consistent practice. That's why it is so important to remember that procedural fluency is more than just memorization and there are many ways to demonstrate it.

A student who instantly recalls 6x7=42 is fluent. So is the student who says, "I know 5x7 is 35, and one more group of seven makes 42." Both approaches show fluency, just in different forms. For students with dyscalculia, flexibility is often the key. **If we define fluency as memorization and recall, we risk making it a gatekeeper to higher level math.** But when students have access to tools and strategies that make sense such as visual models, fact tables, and even calculators, they can solve problems confidently and efficiently. These aren't shortcuts or crutches. They are signs of meaningful, strategy-based fluency rooted in understanding.

Strategic Competence

Strategic competence refers to a student's ability to formulate, represent, and solve mathematical problems. It's not just about getting an answer, it's about understanding what the problem is asking and deciding how to approach it. This strand involves the ability to build a mental representation of a problem, select appropriate tools or strategies, and navigate through tasks to a solution.

In the classroom, many problems are already neatly structured. The question is clear, the numbers are friendly, and the operation is obvious. But outside of school, problems are often messier. Strategic competence prepares students for that uncertainty. It asks them to see how math connects to many types of situations and plan a pathway towards a solution.

For students with dyscalculia, this strand can sometimes reveal strengths that are hidden in more traditional tasks. While they may struggle with memorized facts or multi-digit procedures, they may also thrive when asked to make sense of a real-world scenario or think creatively about how numbers work together in a situation that matters to them. They may not always get the correct answer, but their approach may reveal flexible, strategic thinking, key traits of strong mathematical problem solvers.

One powerful instructional shift that nurtures strategic competence is the use of open-ended word problems. These problems have multiple entry points and more than one possible solution. These problems encourage students to take ownership of how they approach a task, explore various strategies, and articulate their thinking. One example that always leaves an impression with students is YouCubed's Foot Parade: students first list animals and the number of feet they have. They then create a parade totaling 12 feet, using any combination of animals. The task is playful and flexible. It invites choice, experimentation, and sense making. I have even had classes that spent a considerable amount of time debating whether a snail has one foot or none and then modeled their ideas by building their parades.

While working through an open-ended task such as this, students aren't just adding. They are reasoning, representing, and revising, all key aspects of strategic competence. For students with dyscalculia, tasks like these are especially powerful. Even if computation is challenging, their capacity to think flexibly, build models, and decide how to approach a situation can shine brightly here. Open-ended problems make it possible for students to engage meaningfully with math, regardless of their levels of fluency or computation skills.

Another powerful strategy is using numberless word problems, a routine popularized by math educator Brian Bushart. In this approach, a word problem is revealed in stages, without any numbers or questions at first. Students are then asked to discuss the context of the problem and make meaning of what is happening. As the numbers are gradually revealed students revise or build upon their thinking. For students with dyscalculia, this routine reduces cognitive load, supports the development of mental models, and fosters confident problem solving.

Strategic competence reminds us that math is not just about computation, it's about making meaning. When we create opportunities for students to make sense of problems in varied ways, we uncover capabilities that traditional tasks may overlook. For students with dyscalculia, this strand can be a powerful strength, offering a window into their creativity as problem solvers.

Numberless Word Problems

Let's walk through some numberless word problems together. Each step is marked by the teacher revealing parts of a word problem and then asking questions to guide students in making sense of it.

Let's start with a comparative word problem, which can be quite tricky for young learners.

A toad ate some dragonflies. A snake ate more dragonflies than the toad.
> *What are you picturing in your mind when you read this?*
> *How many dragonflies are you picturing each animal eating?*

A toad ate 4 dragonflies. A snake ate more dragonflies than the toad.
> *What changed? What did we learn from this new information?*
> *If the toad ate 4, how many dragonflies could the snake have eaten?*

A toad ate 4 dragonflies. A snake ate 3 more dragonflies than the toad.
> *What changed? What did we learn from this new information?*
> *Does this mean the snake ate 3 dragonflies? How do you know?*
> *What questions could we ask about this situation?*

A toad ate 4 dragonflies. A snake ate 3 more dragonflies than the toad. How many dragonflies did the snake eat?
> *What is the question asking?*
> *How can we show this with objects or pictures to show this?*
> *How can we use numbers and symbols to show this?*

Numberless Word Problems can also be used for more complex multi-step situations. For example:

Cindy Tobeck won the 2016 Safeway Championship Pumpkin Weigh-Off in Half Moon Bay. The runner-up weighed some number of pounds. Her winning pumpkin was heavier than the runner up.

What do you think it takes to grow a pumpkin that heavy?

What is a runner-up?

How much do you think the runner-up pumpkin weighed?

How much do you think Mrs. Tobeck's pumpkin weighed?

Cindy Tobeck won the 2016 Safeway Championship Pumpkin Weigh-Off in Half Moon Bay. The runner-up weighed 1,723 pounds. Her winning pumpkin was heavier than the runner up.

What do we know now that we didn't know before?

How does the weight of the runner-up pumpkin compare to our estimates?

Do you want to revise your estimate for the weight of Mrs. Tobeck's pumpkin?

Cindy Tobeck won the 2016 Safeway Championship Pumpkin Weigh-Off in Half Moon Bay. The runner-up weighed 1,723 pounds. Her winning pumpkin was 183 pounds heavier than the runner up.

What changed? What do we know now that we didn't know before?

Hmm, so is this saying that Mrs. Tobeck's pumpkin only weighed 183 pounds? That doesn't make sense. What questions could we ask about this situation?

Cindy Tobeck won the 2016 Safeway Championship Pumpkin Weigh-Off in Half Moon Bay. The runner-up weighed 1,723 pounds. Her winning pumpkin was 183 pounds heavier than the runner up. Mrs. Toback was paid $6 per pound in prize money.

Act surprised, "Oh, I expected a question, but instead we got more information. What do we know now that we didn't know before?" What does the phrase "$6 per pound" mean?

Do you think Mrs. Tobeck was happy with the amount of prize money she received? $6 isn't a lot of money.

What questions could we ask about this situation now?

Cindy Tobeck won the 2016 Safeway Championship Pumpkin Weigh-Off in Half Moon Bay. The runner-up weighed 1,723 pounds. Her winning pumpkin was 183 pounds heavier than the runner up. Mrs. Toback was paid $6 per pound in prize money. How much prize money did she receive for her winning pumpkin?

In your own words, what is the question asking?

Numberless Word Problems support the development of Strategic Competence by slowing the process down so students can reason through a situation before numbers take over. By revealing information gradually, we help students build meaning, make sense of relationships, and approach the final question with confidence.

Example sequences from Brian Bushart's Numberless Word Problems (https://numberlesswp.com/). Shared with attribution.

Adaptive Reasoning

Adaptive Reasoning is the ability to think logically about mathematical ideas. It includes reflecting on whether a strategy makes sense, justifying why it works, and adapting when it doesn't. Adaptive reasoning involves informal explanation, justification, and intuitive thinking based on patterns, analogies, and prior experience. It's thinking about your thinking, noticing when something feels off, and choosing to try again.

Adaptive reasoning may be one of the most revealing and underappreciated strengths. Students who struggle often come up with their own informal ways of making sense of math. They might talk through ideas out loud, use analogies, or intuitively notice patterns. Even when answers are incorrect, or their recall of procedures break down, their reasoning process can be logical and insightful. They might say things like, "This doesn't make sense because the answer should be smaller not bigger," or "I thought this was the same as what we did yesterday, but it's not." These moments matter. They show that the student is monitoring their thinking and making purposeful decisions about how to move forward. These are core features of adaptive reasoning that don't always show up in written work but are deeply mathematical in nature.

Like other strands, adaptive reasoning doesn't develop overnight. And it looks different across age levels. According to *Adding It Up* (NRC, 2001), "Some researchers have concluded that children's reasoning ability is quite limited until they are about 12 years old. Yet when asked to talk about how they arrived at their solutions to problems, children as young as 4 and 5 display evidence of encoding and inference and are resistant to counter suggestion." This means that even young children, including those with dyscalculia, are capable of reasoning through mathematical ideas when given the chance to explain their thinking in ways that are meaningful to them.

Reasoning In Real Time: A Conversation with Leo

At a table sit Leo and his teacher. On the table sits a closed box and five cubes.

Teacher: There are 9 cubes in this box. And there are some here on the table. We need to figure out how many there are altogether.

Leo: (counting each block on the table) One, two, three, four, five. Five.

Teacher: (gesturing toward the box and cubes on the table) There are five cubes altogether?

Leo: Oh wait. (Looking between the box and the cubes on the table) How many?

Teacher: There are 9 cubes inside this box.

Leo walks over to the nearby math center, takes out 9 cubes, and puts them on the table. He counts all the cubes up beginning at 1, arriving at his answer.

Leo: Fourteen. There're fourteen cubes.

Teacher: How do you know?

Leo: I counted them.

The teacher removes the nine cubes that Leo took from the math center, leaving the closed box and five lose cubes.

Teacher: You did. You counted all fourteen of them. Is there another way you can count them if you couldn't see them all?

Leo: There's nine in the box?

Teacher: Yes.

Leo: (grabbing the loose cubes one at a time) nine, ten, eleven, twelve, thirteen. There's thirteen now? That's not right.

Teacher: Hmmmm. Let's go back. How many are in the box?

Leo: (pauses, thinks) Nine. There's nine in the box.

Teacher: Right. And how many are out here?

Leo: (touches each one slowly) One, two, three, four, five. Five are here.

Teacher: (pointing to the box) What if we start counting the cubes on the table at ten, since we already know that there are nine cubes in the box? The cubes in the box are nine, so then we count on.

Leo: (dragging the box across the table) Nine, (dragging each loose cube) ten, eleven, twelve, thirteen, fourteen.

Teacher: What do you think?

Leo: Yeah. Fourteen. That's what I got before. I counted the cube on the table as nine the first time.

Teacher: So, what helped you?

Leo: You said nine were in the box. So, the cube on the table is ten, not nine.

This conversation between Leo and his teacher illustrates adaptive reasoning in action. Throughout the exchange, Leo demonstrates persistence, flexibility, and a willingness to revise his thinking based on new information. Rather than shutting down after his first misstep, Leo revisits the problem multiple times, testing different strategies and responding to feedback. He builds a mental representation of the situation, self-monitoring when something sounds "not right," and reflects on his earlier error with thoughtful insight. Most importantly, Leo makes a conceptual shift from recounting all the objects beginning at one, to counting on from a known quantity, in this case 9. This marks a meaningful mathematical leap and shows how adaptive reasoning supports learning, even when students struggle with basic skills like counting or memory. Leo's thinking is the perfect example of adaptive reasoning: rich with sense making, adjustment, and justification.

Productive Disposition

Productive disposition refers to a student's belief that mathematics is sensible, useful, and something they are capable of doing. It's more than just attitude. It's about identity. Students with a strong productive disposition see themselves as math learners. They believe that their effort matters, that confusion is part of learning, and that with time and support, they can grow.

According to *Adding It Up* (NRC, 2001), productive disposition develops alongside the other strands. As students build conceptual understanding, procedural fluency, strategic competence, and adaptive reasoning, their confidence grows. But the opposite is also true. When students experience repeated failure, when tasks feel confusing or irrelevant, or when they are consistently asked to perform in ways that don't align with their strengths, their self-perception erodes.

For students who struggle with math, productive disposition is often the most fragile strand. They may begin to see themselves as "not a math person," a label that becomes self-fulfilling over time. Researcher Jo Boaler (2016) emphasizes the importance of disrupting these beliefs by fostering a growth mindset—helping students understand that math ability is not fixed, and that mistakes are essential parts of learning. This is not **just** about being positive. It's about creating the conditions where students experience success and come to believe in their own ability. Productive disposition grows when students feel seen and supported, when they are offered tasks that are accessible but challenging, and when their thinking, not just their answers, is valued.

For students with dyscalculia, this means celebrating strengths, modeling perseverance, and shifting the classroom narrative away from speed and perfection. It means letting students know, explicitly and often, that math is for them, too.

Supporting Productive Disposition: Teacher Moves That Matter

Fostering productive disposition takes intention. Here are a few practices that can help students see themselves as capable math thinkers:

Model a growth mindset:
"This part of the problem really stumped me at first."
"Let's try it a different way and see what happens."

Shift feedback from performance to process:
Instead of "You're so smart" try "You really stuck with that,
even when it got tricky."
"I noticed how you changed your strategy when the first one
didn't work."

Normalize mistakes and confusion:
"Mistakes are a part of learning. What did this one help us
figure out?"
"Can someone build on an idea, or offer a different way?"

Use affirming language and mantras:
"Math is thinking, not just answers."
"Your brain grows when you try something new."

These small shifts in language and tone help shape a classroom
culture where persistence is celebrated, thinking is visible, and every
student feels like math belongs to them.

Strands of Strength: Building A Balanced Student Profile

Understanding a student's struggles is only part of the
picture. To teach effectively we also need to understand their
strengths, not just what they can't do, but what they can. Strengths
can serve as bridges. When a student struggles with fact recall but
demonstrates flexible reasoning, we can build fluency by leaning on
that reasoning. When a student finds written work hard to organize
but explains ideas clearly in conversation, we can use oral language
as an entry point into more abstract representations.

The Strands of Mathematical Proficiency: Conceptual
Understanding, Procedural Fluency, Strategic Competence,
Adaptive Reasoning, and Productive Disposition, offer us a
research-based framework for noticing and documenting these

strengths. They provide specific, observable behaviors that we can look for and nurture, even in students who experience significant math difficulties.

Putting It Into Practice

Let's return to Leo. In the previous chapter, we used the Student Error Pattern Checklist which revealed consistent difficulties with Core Number and Memory. Similarly, using the Strands of Mathematical Proficiency as a checklist can help us uncover patterns of behavior, making it easier for us to identify strengths and make decisions that inform our instruction.

When we shift our lens to notice what Leo can do, a more balanced picture emerges. He may count slowly and rely on manipulatives, but he shows strong reasoning when asked to explain his thinking. He may struggle to retain steps in a multi-step process, but he can model his thinking with objects and revise his approach when prompted. These behaviors are more than just "bright spots". They are actionable indicators of how Leo processes and understands math.

Below is an example of how Leo's strengths might appear when documented using the Strands of Mathematical Proficiency checklist (found in <u>Appendix B</u>).

Student Name: _____ **LEO** _____ Date: ____ **1/16** ____

Grade/Class: ___ **SECOND GRADE** ____

<u>Strands Of Mathematical Proficiency</u>

Directions: Please check off all the behaviors that apply to the student.

Conceptual Understanding
❑ Understands why mathematical ideas are important.
❑ Understands the contexts in which a mathematical idea is used.
☒   Learns new ideas by connecting to what they already know.
☒   Explains the method they use to solve a problem.
❑ Represents mathematical situations in different ways based on different purposes.
☒   Connects what they know to new or unfamiliar problems.

Procedural Fluency

❑ Knows when to add.
❑ Has strategies for addition.
❑ Addition strategies are efficient.
✘ Addition strategies arrive at the correct answer.
❑ Multiplication strategies arrive at the correct answer.
❑ Knows when to divide.
❑ Has strategies for division.
❑ Division strategies are efficient.
❑ Division strategies arrive at the correct answer.

Strategic Competence (Problem Solving)

❑ Formulates a problem that requires math to solve.
❑ Demonstrates understanding of a problem's important information.
❑ Represents a word problem using numbers, pictures, words, diagrams, or graphs.
❑ Identifies which information is unimportant in a problem.
✘ Uses an approach to make sense of a new problem type (using manipulatives, creating drawings, etc).

Adaptive Reasoning

❑ Identifies and explains patterns they see.
✘ Explains their reasoning for an answer.
✘ Determines whether or not a strategy will work.
❑ Abandons strategies that seem ineffective and generates alternative plans.
✘ Justifies whether or not an answer is reasonable.

Productive Disposition

✘ Sees themselves as capable of learning mathematics.
❑ Believes math can be useful.
✘ Shows perseverance.
❑ Has a growth mindset towards mathematics.
❑ Has a positive disposition towards mathematics.

This kind of profile helps us see that Leo has strong Conceptual Understanding and Adaptive Reasoning. He makes sense of ideas by connecting them to what he already knows and can explain his thinking verbally, even when his written work is unclear. These strengths give us a starting point. For example, his ability to explain his reasoning out loud can be used as a bridge to support his memory. By encouraging him to narrate steps aloud as he works, we help reinforce sequences and reduce cognitive overload. His flexible thinking and willingness to revise strategies provide a foundation for developing more efficient procedures, even if fact recall remains slow. Rather than focusing solely on what Leo struggles with, this profile allows us to use what is working as leverage—to build fluency through understanding, to strengthen memory with modeling, and to design instruction that honors his strengths while addressing his needs.

Sofia's Snapshot: Sharing Money

On Sofia's desk a paper reads: You and three friends are sharing $18.92. How much money will each of you get?

Sofia stares at the page. She reads the problem again, lips moving slightly as she writes:

$$\$18.92 \div 4=$$

She begins dividing. 4 goes into 18 four times. 4 goes into 9 two times. 4 goes into 2... zero?

She hesitates. The numbers feel slippery. She writes 420 as the answer anyway and pulls out her calculator to check it, the contents of her math kit spill on her table. She sifts through coins and fraction strips until she finds her calculator. She enters:

$$1892 \div 4 = 473$$

She stares at it. That doesn't feel right either.

Teacher: Hi Sofia. Tell me what you're working on.

Sofia: (Points to the problem) I'm trying to divide. It says we're sharing this money, $18.92, between four people.

Teacher: Okay, walk me through what you've done so far.

Sofia: (Flips her paper to show her scratch work) I tried to divide like normal, 18, then 9, then 2, but it didn't really work. I wrote 420. Then I used the calculator to check it and it said 473. But that doesn't make sense either.

Teacher: Okay. Now what?

Sofia: (Shrugs)

Teacher: Let's take one piece at a time. Where might you start?

Sofia: Maybe the $18?

Teacher: Okay. How can you share $18 with four people?

Sofia: (Whispers and counts on her fingers) Four, eight, (pause to count up in her head) twelve, (pause again) sixteen, twenty… That won't work.

Teacher: Skip counting won't work? What else can you try?

(She places her hands near the money on Sofia's desk, offering a nonverbal prompt.)

Sofia: (Begins counting out 18 one-dollar bills, making four piles) There's two left.

Teacher: How might you share those $2 with four people?

Sofia: Oh! Fifty cents. Because each dollar has 50 and 50 in it.

Teacher: (Begins recording Sofia's thinking on her own paper, drawing four faces) So how much does each person have so far?

Sofia: Fifty cents and… (She picks up one pile of dollar bills and counts) Four. Four dollars.

Teacher: (Draws $4.50 under each face) Now what? Did we share all the money? Let's look back at the problem.

Sofia: We shared the $18. Now we have to share the 92 cents.

(She grabs the teacher's paper and begins writing 10¢ repeatedly under each face) Ten, twenty, thirty, forty, fifty, sixty, seventy, eighty… (Looks up) There's 10 cents left.

Teacher: And two more cents from the 92. So that's 10 plus 2.

Sofia: Twelve cents.

(She draws a 1¢ coin under each face, counting quietly) One, two, three, four… twelve.

Teacher: Let's go back to the question. It says: "You and three friends are sharing $18.92. How much money will each of you get?" Did you share all of the $18.92?

Sofia: (Nods)

Teacher: So where's your answer?

Sofia: (Circles one face and the money she drew underneath it)

Teacher: Perfect. I'm going to check in on some other friends. When I come back, I'll be looking to see how you labeled your work and your final answer.

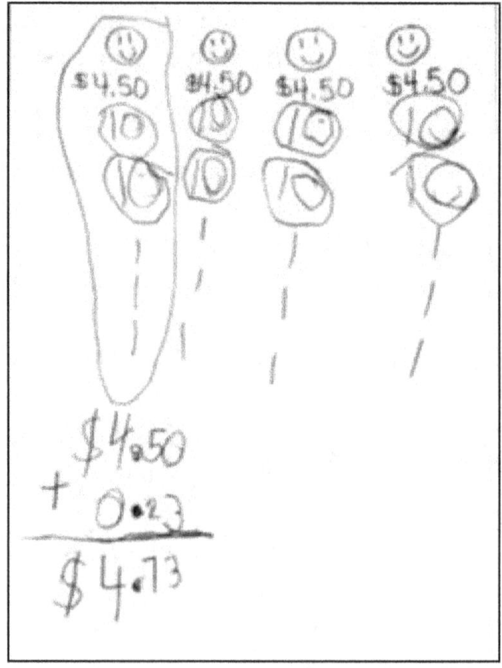

Figure 3.1: Student Work Sharing Money

Note. Student work sample used with permission. Identifying
details have been removed

Interpreting Sofia's Work Through the Strands

Sofia's work on this problem reveals important insights
about how students with dyscalculia may approach multi-step word
problems. Her initial strategy—applying a long division algorithm—
shows a reliance on memorized procedures rather than conceptual
understanding. She makes several calculation errors, and when her
answer doesn't make sense, she turns to a calculator. Even then, the
result feels wrong to her. This moment of uncertainty is critical.
Rather than giving up, Sofia rereads the problem and begins to
reconsider her approach.

With her teacher's support, she shifts strategies. Rather than
relying on an abstract algorithm, she turns to a concrete method:
decomposing the $18.92 into parts, modeling each step with bills,
coins, and drawings. This pivot highlights her ability to engage in
strategic and conceptual thinking when the task is made accessible.

Using the Strands of Mathematical Proficiency, we can identify several areas of strength that should inform instruction and build Sofia's confidence as a capable mathematician.

Student Name: ___Sofia___ Date: _11/16_

Grade/Class: ___5th Grade___

<u>Strands Of Mathematical Proficiency</u>

Directions: Please check off all the behaviors that apply to the student.

Conceptual Understanding
- ❏ Understands why mathematical ideas are important.
- ✘ Understands the contexts in which a mathematical idea is used.
- ✘ Learns new ideas by connecting to what they already know.
- ❏ Explains the method they use to solve a problem.
- ✘ Represents mathematical situations in different ways based on different purposes.
- ❏ Connects what they know to new or unfamiliar problems.

Procedural Fluency
- ❏ Knows when to add.
- ❏ Has strategies for addition.
- ❏ Addition strategies are efficient.
- ❏ Addition strategies arrive at the correct answer.
- ❏ Knows when to subtract.
- ❏ Has strategies for subtraction.
- ❏ Subtraction strategies are efficient.
- ❏ Subtraction strategies arrive at the correct answer.

Grades 3 and Up
- ❏ Knows when to multiply.
- ❏ Has strategies for multiplication.
- ❏ Multiplication strategies are efficient.
- ❏ Multiplication strategies arrive at the correct answer.
- ✘ Knows when to divide.
- ✘ Has strategies for division.
- ❏ Division strategies are efficient.
- ❏ Division strategies arrive at the correct answer.
- ❏ Multiplication strategies arrive at the correct answer.
- ❏ Knows when to divide.
- ❏ Has strategies for division.
- ❏ Division strategies are efficient.
- ❏ Division strategies arrive at the correct answer.

> **Strategic Competence (Problem Solving)**
> ❑ Formulates a problem that requires math to solve.
> ✘ Demonstrates understanding of a problem's important information.
> ✘ Represents a word problem using numbers, pictures, words, diagrams, or graphs.
> ❑ Identifies which information is unimportant in a problem.
> ✘ Uses an approach to make sense of a new problem type (using manipulatives, creating drawings, etc).

Conceptual Understanding

Once scaffolded, Sofia showed that she understands the core idea of division as fair sharing. She successfully broke down the total into manageable parts—first dollars, then cents—and reasoned about how to divide those among four people. She also demonstrated understanding of money equivalence, connecting \$1 to two 50¢ pieces and thinking flexibly with coin values.

Strategic Competence

Although she struggled initially, Sofia demonstrated perseverance and flexibility. She shifted from an ineffective algorithm to a more intuitive strategy using visuals and manipulatives. Her drawn faces, coin groupings, and step-by-step counting reflect meaningful efforts to represent the problem in accessible ways.

Productive Disposition

Sofia stayed engaged throughout the task. Even after encountering multiple breakdowns in memory and reasoning, she continued to try new strategies. Her nod when asked whether all the money had been shared—and the care she took in circling her final answer—suggest a growing sense of ownership over her mathematical thinking.

Let Strengths Lead the Way

When we take the time to examine student thinking through both the lens of dyscalculia subtypes and the Strands of Mathematical Proficiency, we begin to build something more powerful than a list of deficits. We begin to build a profile of the whole child. In this chapter, Leo's and Sofia's stories reminded us that student performance is rarely consistent or clear-cut. Errors may be frequent, but they are not random. And strengths, while sometimes subtle, are always there if we know how to look for them.

By identifying Leo's difficulties with counting and memory, we clarified where support is needed. But it was his persistence, use of tools, and conceptual understanding of number that showed us how to begin. With Sofia, her early reliance on rote procedures and hesitation when those strategies failed pointed to challenges with reasoning and recall. But when she shifted to a concrete approach—using drawings, money manipulatives, and visual groupings—her flexibility, sensemaking, and strategic thinking came to the surface. In both cases, student work became more than a score or a solution. It became evidence of how each learner thinks.

The Strands of Mathematical Proficiency give us a framework to name, nurture, and build upon those strengths. They help us tell a fuller, more hopeful story, one where difficulties are met with understanding and instruction begins not with what's missing, but with what's already working. As we move forward, let this chapter serve as a reminder: our students are not defined by their struggles. They are defined by their thinking. And when we start with strengths, we give every child a way into mathematics.

Reflection Activity for Teams and Book Study Groups
Revisit the student whose profile you began earlier or select a new student from your setting. This time, use the Strands of Mathematical Proficiency Checklist in Appendix B to identify areas of strength.

- Which strands appear to be areas of strength for this student?
- How might these strengths help compensate for the areas of difficulty that were previously identified?
- In what ways would your instructional planning change if you started from a student's strengths rather than their weaknesses?

- How can you make space in your classroom routines to notice and nurture these strengths?

Document your observations and ideas. As we move into later chapters focused on planning instruction, you'll be able to draw on this profile to design supports that are both targeted and affirming.

References

Boaler, J. (2015). *Mathematical Mindsets: Unleashing Students' Potential Through Creative Math, Inspiring Messages and Innovative Teaching.* Jossey-Bass Inc Pub.

Bushart, B. (n.d.). *Numberless word problems* [Website]. https://numberlesswp.com/

Lambert, R. (2024). *Rethinking Disability and Mathematics: A UDL Math Classroom Guide For Grades K-8.* Corwin.

National Research Council. (2001). *Adding it up: Helping children learn mathematics* (J. Kilpatrick, J. Swafford, & B. Findell, Eds.). National Academy Press. https://doi.org/10.17226/9822

Stokke, A. (Host & Producer). (2024, November 8). *How to build automaticity with math facts: A practical guide* [Audio podcast episode]. *Chalk & Talk.* https://www.annastokke.com/ep-36-transcript

YouCubed. (n.d.). *Foot Parade (1–2)* [Lesson plan]. YouCubed. https://www.youcubed.org/wim/foot-parade-1-2/

Chapter 4

Where Are They Now?

By now we have created a clearer picture of our math students. We've gathered information about the types of math thinking that causes breakdowns in understanding and the strengths that we can build from. But there is still one important question left: What does this student understand right now, and how can I help them take their next step? Instead of reteaching what was missed, we're going to identify what already exists in the student's conceptual foundation and use that as a launching point. The tool we will use for this is the CRA Instructional Sequence, a powerful approach that helps us match our instruction to how mathematical understanding develops.

CRA stands for Concrete-Representational-Abstract and is based on the work of psychologist Jerome Seymour Bruner (1966). By observing how a student interacts with hands-on materials, represents their thinking, or how they interact with abstract symbols, we can gain insight into where their understanding is strongest. When working with students who struggle with mathematics, this information is crucial.

Best of all, the CRA instructional Sequence doesn't require you to reinvent your curriculum. Because let's be honest, when you have a class of 30 to teach and are expected to follow the curriculum "with fidelity", that kind of overhaul is just not realistic. Instead, it offers a way to align already existing content to the student's developmental level of understanding. This makes it easier to fine tune lessons to meet the students where they are while still moving them toward grade level standards.

In this chapter we will revisit Leo and Sofia's profiles and look at their thinking through the CRA lens. You'll see how a few small shifts can honor their unique paths to understanding.

What Is The CRA Instructional Sequence?

The CRA Instruction Sequence has been referred to by a variety of names: the CSA intervention, the CRA Framework, the CRA Method. Regardless of its name, it refers to three stages of learning that students move through as they work toward conceptual understanding.

<u>C</u>oncrete. In the concrete stage, students use hands on tools to explore a concept. Cubes, counters, base-10 blocks, fraction tiles, Cuisenaire rods, and pattern blocks are all manipulatives that are typically found in an elementary classroom. At this stage, concepts are modeled and explored through action: joining, separating, grouping, comparing, in ways that connect to real, tangible experiences. This stage builds the foundational understanding students need to make sense of more abstract representations.

<u>R</u>epresentational. (sometimes called pictorial or semi-concrete). In the representational stage, thinking is modeled in written form through the use of pictures, number lines, charts, tables, and arrays. At this stage, students rely less on physical manipulatives and more on the visualization of the concepts involved. This stage acts as a bridge between concrete experience and abstract notation, helping students connect their hands-on understanding to formal mathematical writing.

<u>A</u>bstract. The abstract stage is where students use numbers and symbols to show their thinking. This is usually where textbooks and assessments begin, jumping straight into equations and algorithms. For some students, especially those with strong conceptual understanding, this symbolic language is a flexible, efficient way to express ideas. But for others, especially those with dyscalculia, starting here can feel confusing and disconnected. They may follow steps without really understanding why they work. That's why the CRA sequence is so powerful. It helps us make sure that students aren't just memorizing procedures but making meaning from them.

This table, adapted from the What Works Clearinghouse, outlines big-picture recommendations for supporting students who struggle with mathematics. It doesn't list every concept or tool you might use, but it offers a framework you can adapt to your own classroom and curriculum.

Table 4.1: Common Concrete and Representational Tools

Concept	Concrete	Representational
Counting & Operations	Base 10 blocks Connecting cubes Two-colored counters One-inch tiles Cuisenaire rods	Hundreds chart 5 frames 10 frames Double 10 frames Number line Arrays
Place Value (including decimals)	Base 10 blocks One-inch tiles Connecting cubes Decimal squares	Place value chart Hundreds chart Pictures showing place value
Fractions	Cuisenaire rods Pattern blocks Fraction strips, tiles, or circles One-inch tiles Connecting cubes	Table Number line Strip diagram Pictures showing proportion

Note. Adapted from Assisting students struggling with mathematics: Intervention in the elementary grades (p. 23), by U.S. Department of Education, Institute of Education Sciences, What Works Clearinghouse, 2021, National Center for Education Evaluation and Regional Assistance. https://ies.ed.gov/ncee/WWC/PracticeGuide/26

The CRA Instructional Sequence is sometimes misunderstood as a strategy reserved for young children or students who struggle. In reality, CRA is a developmental pathway that supports all learners in building deep, connected understanding. While the focus of this book is students with dyscalculia and math-based learning differences, planning through the lens of CRA benefits every student. By aligning instruction with how understanding develops, not just what the pacing guide says, we make math more meaningful and accessible.

The stages, concrete-representational-abstract, are not rigid levels to move through only once in a fixed order. Instead, they offer multiple entry points for students to access mathematical ideas and reinforce their learning. For students with dyscalculia, who often need more time and support to internalize number relationships and procedures, CRA offers a powerful structure for helping students see math as clear and purposeful.

CRA In Action: Leo

As you may recall, Leo is a second-grade student who experiences persistent difficulty with counting, representing quantities, recalling addition facts and keeping track of steps. Our Student Error Pattern Checklist in chapter 2 revealed that Leo demonstrates behaviors that are consistent with the core number and memory subtypes of dyscalculia. However, in chapter 3 we discovered that Leo has significant strengths in conceptual understanding and adaptive reasoning. He connects new concepts to what he already knows, explains why a method works, and justifies his solution, especially when given time and tools that support his thinking.

It's now January and within the next two weeks, Leo's class will begin a unit of study on place value. To better inform our instructional steps, let's examine his current level of understanding across the stages of the CRA Instructional Sequence. This isn't a formal assessment or an elaborate protocol, just a brief conversation and a simple task to help us see how Leo thinks. Because let's be honest, no one has time for fancy. We just need something that works.

Concrete Stage

"Leo, please show me 18 with the base ten blocks.," his teacher asked. Leo immediately counted 18 individual cubes out loud, "One…two…three…" Proudly announcing "Eighteen!" when done.

"Now add six please," his teacher requested. Leo looked at his hands and counted six of his fingers. He held them up close to his face and whispered, "Eighteen." Then tapped each of his fingers to his chin as he counted on, "Nineteen, twenty, twenty-one, twenty-two, twenty-three, twenty-four. Twenty-four!" He declared.

Strengths Highlighted:

- **He shows one-to-one correspondence:** Leo carefully matched each cube to a number as he counted, something he used to skip or rush through. This shows he's building stronger coordination between number words and actual quantities.

- **He knows number sequences:** Leo was able to count all the way to 24 without hesitation. He didn't just say the numbers, he understood that twenty-four was the total after adding six more to eighteen.

- **He understands what addition is:** When asked to add six, Leo didn't guess or start over. He knew to count on from eighteen. This shows he understands addition as moving forward from a number.

- **He uses tools that work for him:** Leo naturally turned to his fingers and whispered number words to himself, a great example of how he uses personal tools and self-talk to stay focused and work through a problem.

- **He's proud of what he knows:** Leo's big smile and enthusiastic "Twenty-four!" shows how proud he felt of his answer.

Representational Stage

After solving the problem with cubes and fingers, Leo's teacher took out a pencil and paper. "Can you show what you did on this paper, please?" She asked. Leo picked up the pencil and began to draw two hands, each with fingers clearly outlined. He carefully labeled each finger with a number, stopping at 6. When asked to explain his drawing, Leo said, "I used my fingers to count on. I started at eighteen and went up six."

Strengths Highlighted

- He can recall and represent his strategy: Leo's drawing shows that he's able to think back on what he did and capture it in pictures. That ability to make his thinking visible is a powerful bridge between the concrete stage and his budding understanding of the representational stage.

- He explains his process: By labeling each finger, Leo links his physical counting to a drawn model, helping to solidify his understanding of addition as a forward moving action, not just a rote task.

Even though his drawing was simple, it captured the essential steps of his thinking. With support, Leo can begin to explore other visual models, like number lines or bar models, that serve the same purpose.

Abstract Stage

Finally, Leo was asked to write a number sentence to match his drawing. He stared at the page for a moment, then carefully wrote 816.

When asked to explain, he pointed to the "81" and said, "I had 18," then pointed to the 6, "and that's how many I added." His teacher recognized a familiar error, one that appeared in earlier work. Leo was switching the digits of 18 into 81, a reversal noted during his initial assessment in Chapter 2.

Strengths Highlighted

- **He attempts to connect symbols to meaning:** Even though his notation was incorrect, Leo clearly understood that the number sentence should reflect what he did with cubes and fingers. His verbal explanation showed stronger understanding than his written response.

- **He is developing symbolic awareness**: Leo knew that he needed to represent both the starting number and the amount added. With support, he's ready to practice connecting those parts to the correct notation (e.g., 18 + 6 = 24).

- **His errors are consistent:** Leo has shown a predictable error of switching digits when writing two-digit numbers. Because it's predictable, it shows a clear pattern in his thinking which gives a clear direction for targeted instruction.

Figure 4.1: Image of Student Work Adding 18 + 6

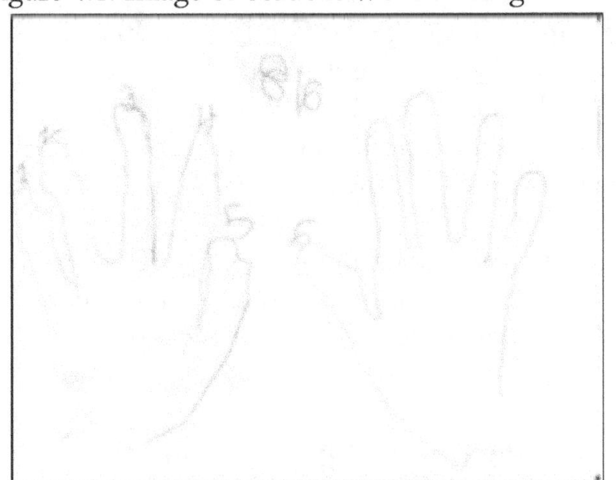

Note. Student work sample used with permission.
Identifying details have been removed

Planning Instruction Based On CRA

Once you've observed a student working across the CRA Instructional Sequence, you're equipped with information about where their understanding is anchored and where it can grow. This is powerful information we can use to plan instruction that targets exactly what needs attention to move forward.

Here are some ways CRA-informed observation can inform planning:

- Pinpoint misconceptions. Watching how students model, represent, or write their thinking can reveal where understanding breaks down. For example, infusing place value blocks or reversing digits in written numbers signals a need to revisit how tens and ones are built and named.

- Choose meaningful entry points. CRA helps you decide where to begin instruction. A student working in the abstract stage may not need remediation, but they may need opportunities to revisit a concept concretely or visually.

- Set appropriate, achievable goals. Once you know what a student can do at one stage, the next goal is usually to help them connect that understanding across another stage.

Sample Language To Guide Your Planning

"If a student can demonstrate the concept using manipulatives, the next step is to help them model it on paper."

"If they can represent their thinking in pictures, the next step is to give explicit instruction on how to accurately notate it with equations."

"If they're writing equations, but cannot explain what they mean, the student needs opportunities to revisit the concrete or representational stages in order to build connections."

Putting It All Together

Leo's work across the CRA Stages gives us a much clearer picture of what he understands and where he will need support in the upcoming place value unit.

At the concrete stage, he confidently counts individual cubes and uses self-selected strategies, showing one-to-one correspondence and a developing understanding of addition. To build the foundation for place value, Leo will need to begin unitizing, grouping objects into tens. Introducing manipulatives such as ten frames, bundled craft sticks, or base-ten blocks will encourage this shift from counting by ones to thinking in groups of ten.

In the representational stage, Leo was able to explain and label his thinking with simple drawings that mirrored his concrete strategy of counting fingers. As he grows more comfortable building two-digit numbers with grouped manipulatives, Leo can be encouraged to represent those tools in his drawings. Moving from fingers to drawn ten frames or bundles will help him see visual representations as symbolic models of ideas, not just memories of actions.

When he moved to the abstract stage, Leo's thinking was still evident, but his number writing revealed lingering misconceptions, particularly with switching digits and omitting operation symbols. We can support Leo by providing explicit instruction on forming two-digit numbers and offering targeted games even outside of the typical math time to give him additional opportunities for practice. Reinforcing the connection between symbolic notation and visual models can help solidify his understanding and reduce confusion.

This progression confirms that Leo has a growing conceptual understanding of counting and addition but needs more time to connect his strategies and representations to conventional math notation. With thoughtful planning and the right tools, we can support Leo to make meaningful progress.

CRA Goal Setting Tool

Name: _Leo_
Date: _1/10_

Task: $18 + 6$

	Strengths Highlighted	**Possible Goals**	**Planning Implications**
Concrete Stage	• one-to-one correspondence • number sequence to 24 • acts out addition • uses blocks (single cubes only) and fingers • shows enthusiasm	Build two-digit numbers in groups of ten and loose ones	Manipulatives that encourage unitizing ten • ten frames • bundles of craft sticks • base ten blocks
Representational Stage	• Draws fingers to represent strategy • Labels fingers with numbers to represent adding • Explains his picture	Represent two-digit numbers with groups of ten and loose ones	Encourage pictures that represent the manipulative he uses
Abstract Stage	• Attempts to include numbers in a number sentence • Attempts to represent starting number and amount added • Consistently switches digits in two-digit numbers	• Accurately record two-digit numbers • Accurately write addition number sentences.	• Explicit instruction on forming two-digit numbers • Additional practice through games

CRA In Action: Sofia

Now let's return to Sofia, a fifth grader whose learning profile offers a different set of strengths and needs. In Chapter 2, we saw her struggle with tasks that required memory and reasoning. She lost her place in multi-step procedures, misapplied algorithms, and needed frequent prompting to move forward. But in Chapter 3, her strengths came into sharper focus. Sofia demonstrated strong conceptual understanding and strategic competence. She made sense of math through context, used visual models to organize her thinking, and found creative ways to represent and solve problems, especially when given the space to explore.

In chapter 3, we observed Sofia attempting to solve a real-world division problem: You and three friends are sharing $18.92. How much money will each of you get? At first, she defaulted to a written algorithm, dividing digits one at a time, as she had likely been taught. But procedural memory didn't hold. Her partial quotients were inaccurate, and her answer, 420, reflected confusion rather than understanding. Even after checking her work with a calculator, she was still unsure of what the solution meant. She knew something was off but couldn't reason through it.

That moment of pause, that sense that "something doesn't feel right", became a critical opening. With teacher support, Sofia revisited the problem using a more grounded approach. Her teacher prompted her to begin concretely: working with coins and bills to model $18.92 and distribute it equally among four people. From there, she moved into the representational stage, drawing money amounts beneath each person's face to track the distribution visually. Only after Sofia had made sense of the problem with concrete materials and visual representations did her teacher suggest adding labels, an invitation to more abstract notation.

Sofia's thinking illuminated something important: she doesn't struggle because she doesn't understand math. She struggles because abstract symbols outpace her reasoning and memory. When the focus shifted to what she could see, touch, and draw, Sofia demonstrated strong strategic competence, sense-making, and perseverance. In that moment, CRA helped her find clarity not by simplifying the math, but by making the math visible.

Instructional Implications

For students like Sofia, starting at the abstract level (as many curriculum materials do) can lead to early breakdowns. The CRA sequence provides a scaffold for re-engaging students by anchoring math in real-world meaning first. In future lessons on division with decimals, Sofia would benefit from:

- Concrete tools: Bills and coins

- Representational strategies: Drawing models, open number lines, tables, or part-whole diagrams.

- Verbal reasoning: Encouraging Sofia to explain her actions aloud as she models and records her steps.

- Delayed abstraction: Waiting to introduce numerical notation and formal algorithms until she has demonstrated understanding with tools and pictures.

This doesn't mean avoiding formal math entirely—it means arriving there with confidence and understanding. So, let's record Sofia's goals and add it to her student profile.

CRA Goal Setting Tool

Name: _Sofia_

Date: _1/12_

Task: $18.92 / 4

	Strengths Highlighted	Possible Goals	Planning Implications
Concrete Stage	Uses manipulatives when prompted.	Choose manipulatives as a tool to solve a problem, or to create meaning when understanding breaks down.	Establish routines for regularly using manipulatives.
Representational Stage	Teacher began scribing Sofia's strategy. Sofia completed the picture sharing 10's and 1's across 4 people and circled her answer.	Represent thinking using pictures, numbers, or words on a page.	Include worked examples of tasks for her to use as models.

	Strengths Highlighted	Possible Goals	Planning Implications
Abstract Stage	Attempts partial quotient with confusion.	Label representational work with number sentences.	Include worked examples of tasks for her to use as models

Tips For Using the CRA Instructional Sequence

Start with what students already know. When introducing a new task or concept, begin by anchoring it in something familiar. This might mean inviting students to use manipulatives, act out a scenario, or build a model before diving into the formal lesson. For example, before solving problems about perimeter, ask students to build shapes with tiles or sticks and walk the outline with their fingers. These tactile, visual experiences give students a mental model to draw on when numbers and symbols are introduced. This is especially important for students with dyscalculia, who will need more concrete experiences to internalize relationships between quantities.

Yes, even in the upper grades. Older students might resist using manipulatives at first, worried that it seems "babyish" or that they'll look like they don't get it. But when we normalize these tools as part of mathematical thinking, not a sign of struggle, we give all students permission to access ideas in ways that make sense to them. Modeling, drawing, and building are not just stepping stones; they are valid ways of thinking, and they belong in every classroom.

Keep grade level goals in focus. Adapting a task to meet students at their CRA stage doesn't mean watering down the math. Instead, it means providing multiple pathways to the same destination. Students can still work toward grade level standards, even if their entry point looks different. For example, a student working on multiplication might model equal groups with counters while others use equations. Both are accessing the same concept, but one is doing it through a concrete lens. The key is to hold the rigor while adjusting the route.

Allow students to show understanding in different ways. Instead of insisting that every student represent their work with only numbers and symbols, invite a variety of approaches. Can they show it with a drawing? Build it? Explain it to a partner? Write it in words? When we value diverse ways of showing thinking, we make space for students who process math differently, and we gain richer insight into what they really understand. For some learners, their understanding may shine more clearly in a picture or a model than on a worksheet.

Use bridging questions to connect stages. One of the most powerful teacher moves is to ask questions that help students connect concrete, representational, and abstract thinking. These "bridging questions" don't just prompt answers. They prompt reflection. For example, after solving with manipulatives, you might ask, "Can you draw what you just did with the counters?" Or "How would you write that as an equation?" These questions help students internalize the idea that math is not about isolated steps but about relationships between actions, visuals, and symbols. Over time, these bridges can create lasting understanding.

Slow down. CRA is not a straight path. Students cycle through stages in different ways, often returning to earlier representations as new concepts are introduced. This flexibility builds lasting understanding. Resist the urge to put the manipulatives away too soon. Give students time to explore and develop understanding at the concrete and representational stages.

Less is more. Avoid overwhelming students by introducing too many models at once. Switching from number lines to ten-frames to bar models can confuse rather than clarify. Repeating 1-2 manipulatives or visual models across topics helps students make connections and reduces the cognitive load of learning new tools. Sticking with familiar, effective tools helps students focus on the math, not the model.

When we slow down long enough to observe how our students interact with materials, pictures, and numbers, we begin to understand the story behind their struggles and their strengths. For students like Leo and Sofia, this approach makes space for real thinking to unfold, not just answer-getting.

The CRA Instructional Sequence isn't a magic fix. It won't erase every challenge or eliminate the need for support. But it gives us a roadmap that starts with what the student can do and builds from there. It helps us bridge the gap between concrete experience and abstract reasoning in a way that feels clear, connected, and doable. And best of all, it doesn't require a new curriculum, a new program, or a complete reinvention. Just a shift in how we look, listen, and respond.

So, as you plan your next lesson, pull a small group, or pause to confer with a student midtask, ask yourself: Where in the CRA sequence is this student working right now? What small move might help them take their next step?

You don't need fancy.
You just need something that works.
And *this*?
This works.

Reflection Activities for Teams and Book Study Groups Exploring the CRA Instructional Sequence

Use these prompts and activities to guide collaborative reflection and planning:

1. Where Do Your Students Land?

Think about a student in your setting. What stage of the CRA sequence do they seem most confident in? Where do they need support?

2. Revisit A Task

Choose a recent math task from your curriculum. How might that task be modeled in each CRA stage? What changes would make it more accessible to a student working concretely? How would you support them moving forward?

3. Inventory and Prioritize Tools

Skim through your current curriculum materials. Which manipulatives and visual models are already built in? Which ones appear most often across units? As a team, identify 2-3 core

---

tools or models you want to prioritize this year. How might consistency in tools help reduce cognitive load and deepen understanding?

4. **Adapt With Intention**

In small groups, pick a core math concept (e.g., fractions, place value, multiplication). Plan out one activity at each CRA stage. Then ask: How can we connect these stages so students see the math ideas, not just the tools or steps in an algorithm?

5. **Use Student Work As A Window**

Bring in a piece of student work. What does the work tell you about the student's understanding? What questions could you ask, or supports could you offer, to help them grow. Use the CRA Goal Setting tool in Appendix C to record your thinking.

6. **Commit To One Small Shift**

As a group, reflect on your current instruction. What's one thing you might try this week to slow down, scaffold, or strengthen the bridge between representations? Write it down and revisit it at your next meeting?

References

Bruner, J.S. (1966). *Toward a Theory of Instruction*. Harvard University Press.

Pennsylvania Training and Technical Assistance Network. (n.d.). *Concrete-representational-abstract: Instructional sequence for mathematics* [PDF]. PaTTAN. https://www.pattan.net/getmedia/9059e5f07edc-4391-8c8e-ebaf8c3c95d6/CRA_Methods0117

U.S. Department of Education, Institute of Education Sciences, What Works Clearinghouse. (2021). *Assisting students struggling with mathematics: Intervention in the elementary grades* (WWC 2021003). National Center for Education Evaluation and Regional Assistance. https://ies.ed.gov/ncee/wwc/PracticeGuide/21

Chapter 5 Planning For Instruction

What we *can* do is make intentional, high-leverage instructional moves

Even with the clearest picture of student needs; strengths, subtypes, CRA stages; we're still left with a big question: How do we teach grade-level content to students who are one, two, or even more years behind?

The reality in most general education classrooms is this: We're already 12 days behind the pacing guide, which means we'll have to consolidate a few lessons to catch up… but the unit assessment focuses on a standard most of the class still needs more time with… but we really need to start the next unit if we want to finish before state testing. Meanwhile, some students haven't mastered the standards from two years ago. And that's exactly when you hear your colleague down the hall say the principal just started another round of surprise observations.

We reach for familiar buzzwords: differentiation, accommodation, and modification. They're everywhere. They're what we're told to do. But what do they really mean in practice, and more importantly, will they actually help students who are significantly behind?

The truth is the gap between where these students are and where the curriculum expects them to be can feel overwhelming. It can seem like they need something entirely different. And in some situations, they might. But in ours, we can't always scrap the lesson plan and start over. We're working within real constraints: time, curriculum, and class size to name a few.

What we *can* do is make intentional, high-leverage instructional moves that increase access, deepen understanding, and help students make progress, without needing to reinvent the wheel.

In this chapter, we'll explore six powerful strategies that have been proven effective by experts in math education and institutions like the What Works Clearinghouse. These aren't bonus activities or intervention tricks. They're just good teaching, effective for all learners, but *essential* for students with dyscalculia. They are:

- Mathematical Language
- Manipulatives
- Worked Examples
- Think Alouds
- Targeted Feedback, and
- Games

It's easy to see these six strategies as just another list of "best practices". But when we look closer, each one offers a tangible way to bridge the gap between conceptual understanding and procedural fluency. Each of these six strategies supports access to grade-level mathematics in a specific way. Together they create a framework for instruction that meets students where they are without lowering expectations. We will begin with the first, and perhaps most essential, strategy: mathematical language.

Mathematical Language: More Than Just Words

Mathematical language is more than just vocabulary. It's a precision tool used by students to understand and communicate mathematics. While everyday conversational language helps us connect with others, it often falls short when conveying exact mathematical meaning. For example, the casual phrase "take away" may work when talking about removing 4 cookies from a plate of 6, but it doesn't connect to other subtraction situations such as comparing whose plate has more cookies. In contrast, the word subtract communicates the mathematical idea in a way that is consistent, precise, and transferable across different mathematical contexts.

Explicit attention to mathematical language benefits all students, but it's especially important for those who struggle. Using the same terms across classrooms and settings helps students connect new learning to grade-level expectations, where precision in communication is increasingly emphasized. Research shows that students need to encounter mathematical vocabulary many times, in meaningful contexts, before it becomes part of their everyday thinking.

Putting It Into Practice

The *Assisting Students Struggling with Mathematics: Intervention In The Elementary Grades* guide (What Works Clearinghouse, 2021) outlines three practical ways to strengthen mathematical language.

1. Make vocabulary instruction explicit and routine.

Introduce terms using student-friendly definitions and a variety of examples: concrete, representational, and abstract. For concrete experiences use manipulatives like counters, base-ten blocks, or fraction strips, to model what a word means. Representational examples may include drawings, number lines, tables, or diagrams. These should directly connect to the manipulatives you are using. Abstract examples show the symbolic notation of the term, like an equation or fraction. Role play is another powerful tool. Students can act out a scenario that demonstrates a concept such as "sharing" objects for division or "grouping" items for multiplication. Using a variety of experiences ensures that students encounter the vocabulary in ways that are meaningful, memorable, and transferable.

2. Use accurate mathematical terms yourself.

Mathematical vocabulary is something we build together in the classroom, and it starts with the words we choose as teachers. Selecting words to teach does not need to be overwhelming. State standards and curriculum guides can support you in identifying terms that matter most. Model precise language in your lessons and everyday conversations, so students can hear and practice it often. When a term has more than one meaning, like a table as furniture versus a data display, pause to point out the difference. Keep vocabulary consistent across lessons, homework, and family communication. This intentional clarity gives students the foundation to grow in confidence and understanding.

3. Support student use of language.

One of the most powerful ways to help students internalize mathematical vocabulary is to give them frequent opportunities to

use it themselves. Encourage students to explain their thinking both aloud and in writing, so they can practice connecting words to ideas. Providing sentence starters such as "The sum is _____ because _____," or guiding questions such as "Can you explain your first step and why you chose it," can help scaffold their explanation while reinforcing correct terminology. When a student uses imprecise wording, gently restate their response using the correct vocabulary, modeling how to communicate clearly without interrupting the flow of thinking. Finally, display key terms prominently in the classroom on a word wall so students can reference them often.
This makes the language visible, accessible tool for reasoning and problem solving.

Teaching vocabulary doesn't need to be an "extra" lesson complete with its own activities. Embed it into the math you are already teaching so that every problem solving discussion is an opportunity to have a language-rich experience. When mathematical language support is purposefully and naturally woven into instruction, it becomes a tool for thinking not just a list of words to memorize.

Building a Math Word Wall That Works

A well-designed math word wall isn't just decorative. It's a living reference tool that supports comprehension, memory, and independence. In my classroom, I have found that students with dyscalculia benefit from having consistent visual and verbal cues to anchor mathematical terms, especially those that are abstract, have multiple meanings, or are easy to confuse.

Here's What I Learned:

❑ Include visuals and examples. Don't just list the word. Add a picture, number sentence, or diagram
For example, next to "sum" include 4+3=7 and a bar model showing parts and whole.

❑ Use student friendly definitions. Keep language clear, simple, and relatable.
Replace "a value resulting from addition" with "the answer when you add numbers together".

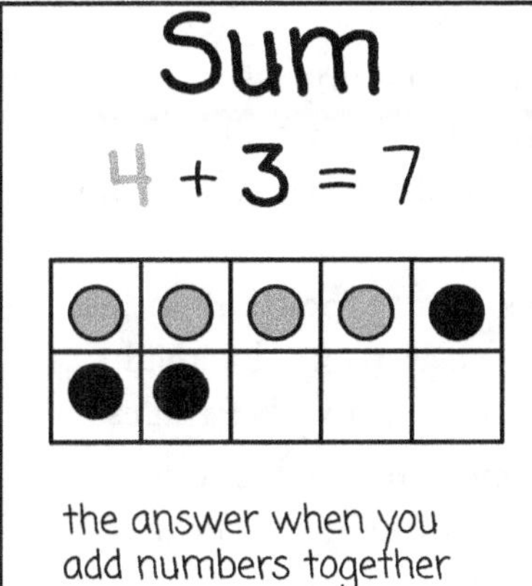

❑ Highlight everyday math words. Words like more, less, each, and altogether can be tricky. Add color coding or icons to help students distinguish between them.

❑ Organize by concept, not alphabet. Group words under headings like Addition & Subtraction, or Fractions. This makes it easier for students to find and use terms meaningfully.

❑ Add words over time. Add words to your word wall as it comes up in instruction, not all at once. Revisit and refer to the word wall regularly to keep it relevant and alive.

Manipulatives

For many students, concrete manipulatives and semi-concrete representations are a temporary step on the path to abstract thinking. But for students with dyscalculia, however, these tools may be needed more consistently and long-term. As Dr Honora Wall notes in *Teaching Students with Dyscalculia* (2022), "Children with a weak foundation can receive remediated instruction and move on successfully without support. Children with dyscalculia will always need support, as dyscalculia persists throughout the lifespan." This doesn't mean using manipulatives forever but rather being strategic about which representations are most effective for each student.

While chapter 4 explored the CRA sequence in detail as a means to identify where a student's understanding of a particular concept lies, it also plays an important role in planning for instruction. Our goal is to provide the supports students need to access grade level mathematics with understanding, without assuming they can, or should, fully transition to abstract notation at the same pace as their peers.

Choosing Representations

- Use standards and curriculum guides as a compass. Prioritize manipulatives and representations that align closely with grade-level expectations and the concepts being addressed.

- Maintain consistency across settings. The same tools and visual supports should be used in the classroom, in small groups interventions, and in communication with families.

- Connect to abstract notation. Whatever manipulative or representation you choose, make sure students see how it links to symbols, equations, or formal procedures. Manipulatives and representations are bridges to understanding abstract ideas.

Putting It Into Practice

Observe your student's interactions with different representations. Which tools help them make sense of a concept? Which facilitate progress toward abstraction? Use this information to plan long-term supports, including manipulatives, visual models, and reference aids. The goal is to increase access to grade-level mathematics while providing consistent, meaningful scaffolds that can remain in place as needed.

From Temporary Scaffold to Ongoing Need

Many classroom supports are intended for short-term use while students learn new concepts and skills. But Dr. Wall points out that students with dyscalculia will need ongoing supports throughout their lives. So, if you notice a student who seems "stuck" in a CRA stage, it's not a sign of poor teaching. It can be a clue into the student's learning profile.

If a student relies on scaffolds far longer than expected and they will lose access to grade level content without it, it could be a sign to consult with the school-based support team and collaborate with the family. In some cases, long term use of a support, like a calculator, number line, or manipulative, may be appropriate. But remember, certain standardized tests only allow these tools if they are a part of a student's IEP or 504 plan.

If you notice that a student's participation or success in math consistently depends on a particular support, that's your cue to:

- Document patterns in your observations and student work.
- Consult the school-based support team to explore targeted intervention options.
- Partner with the family to share data, discuss concerns, and consider next steps.

And most importantly, avoid saying things such as, "I think this child has dyscalculia." This can be interpreted as an attempt to diagnose. Not all students who struggle with math have dyscalculia, and determining that is outside our scope. Our role is to notice patterns, document what we see, and collaborate with others to ensure that students get the right help at the right time.

Worked Examples

A worked example is a fully solved problem that students can study and reference while learning to solve similar problems on their own. Instead of diving straight into independent practice, students first see the thinking process and solution steps clearly laid out.

Maria Chiara Fastame (2021) notes that one of the most evident difficulties for students with dyscalculia is the inability to generalize; to take what they have learned in one context and apply it to another. To address this, she suggests that students be provided with a wide range of problems. She suggests that varying contexts and representations can build flexible thinking.

Worked examples are not intended to be used as a step-by-step guide to mimicking an algorithm without conceptual understanding. When implemented thoughtfully, worked examples reduce cognitive load, highlight strategic thinking, and make it easier for students to see patterns across problems. Over time, they shift from being a tool to a mental model that students can apply to new situations independently.

Worked examples don't need to be a separate lesson. In fact, one of the most powerful approaches is to revisit problems solved together in previous lessons. These shared experiences become anchor points students can return to, especially if they have struggled with similar problems in the past. Posting solved problems on an anchor chart or keeping them in a math book or reference folder makes them an ongoing tool for learning.

Think Alouds

Teachers are often trained to model their thinking out loud, showing students how to approach a problem step-by-step. This "teacher think aloud" offers a window into the strategies, decisions, and even mistakes that experienced learners make along the way.

Maria Chiara Fastame (2021) suggests taking the process a step further by having students be the ones to verbalize their own strategies as they work. This practice encourages students to slow down, reflect, and make their reasoning explicit. It also helps prevent the common habit of solving problems impulsively by randomly combining numbers.

When students articulate their thinking, teachers can immediately identify misconceptions and address them before they become engrained. As the saying goes, "perfect practice makes perfect." If students are practicing incorrectly, silently reinforcing errors, they risk strengthening the wrong pathways. Think alouds create an opportunity to guide learners toward accurate, efficient strategies while fostering metacognition.

Of course, it isn't realistic to sit beside every student during every lesson to hear their full thought process. One way to make this work in a busy classroom is to strategically incorporate think aloud into partner work expectations. Providing sentence stems such as "First, I…," "I decided to…because…," or "I changed my mind when…," can give structure to the conversation and keep the focus on reasoning rather than just answers. As part of this structure, make it clear that partners should listen actively and respectfully challenge each other's reasoning, asking questions like, "Why did you choose this step?" or "I don't understand this part of your strategy. Can you explain it again?" Encouraging this kind of peer questioning helps students refine their strategies and learn from one another. Partner work allows every student to verbalize their thinking, while the teacher circulates to listen in and provide targeted feedback where it's most needed.

Targeted Feedback

Maria Chiara Fastame (2021) distinguishes between two main types of feedback. Extrinsic feedback tells students whether they did well or not overall, often in the form of a grade, score, or simple "right/wrong". Intrinsic feedback goes deeper, providing specific information about the actions a student took toward a goal and how those actions can be refined. Both types are necessary at different times. **Extrinsic feedback can boost motivation and celebrate milestones, while intrinsic feedback drives learning forward by highlighting what to adjust and why**.

Targeted feedback can take many forms: spoken or written, from teachers, peers, or family at home. When the expectation to give constructive, respectful feedback is made routine, it not only helps the recipient but also sharpens the giver's own understanding. Games, whether in person or online, can be powerful tools for delivering this kind of feedback. They create low stress, engaging opportunities where students can immediately see the impact of their decisions, adjust strategies, and try again in real time.

Games naturally create built-in feedback loops, making them an ideal way to provide both extrinsic and intrinsic feedback without adding extra grading or conferencing time. In a math game, extrinsic feedback might come from a score, winning a round, or moving ahead on a game board as clear indicators of performance that can be motivating and energizing. Intrinsic feedback happens in the moment, as students see how their decisions affect the outcome. For example, realizing that a miscalculation caused them to lose points prompts immediate self-correction and strategy adjustment.

To make feedback more powerful, choose games that:
- Allow for multiple attempts so students can quickly apply adjustments.
- Encourage discussion between turns, so players can explain moves and offer suggestions.
- Provide a balance of chance and skill, reducing pressure while still rewarding sound strategies.
- Discourage speed in computation.

Digital games can enhance this further by giving instant, specific feedback (e.g., highlighting an error and showing the correct step), while physical board or card games invite verbal feedback from peers. Whether online or in person, pairing students strategically, matching different strengths, can ensure each player benefits from peer input. Over time, these game-based experiences normalize feedback as a natural, positive part of learning rather than a judgment.

Games also make excellent homework options. They bring families into the learning process, offering a fun, low-stress way to review skills at home. Because games can be tailored to target a specific student's needs, whether that's a need from the current grade level, or a previous one, they allow for meaningful practice that feels more like play than work. This combination of engagement, personalization, and feedback makes them a powerful extension of classroom learning.

Games

Learners with dyscalculia need a lot more repetition, spread out over a longer period of time, than their peers. Relying on rote practice through worksheets or recitation can quickly become boring. Ronit Bird, in *The Dyscalculia Toolkit*, suggests that the use of games can deliver more practice in ways that are interactive, enjoyable, and motivating.

According to Bird, a well-designed math game targets a single mathematical idea at a time and uses concrete materials to make the concept tangible. Unlike worksheets, which often allow for only mechanical practice, games require players to go through the logical and deductive steps needed to reach an answer. This ensures students are not just arriving at a result but are actively reasoning their way there.

Students are naturally drawn to games. They rarely notice how much learning and reinforcement is happening because they're focused on the interaction and fun. The key is intentional design: games should not be treated as an "extra", or a time-filler, and they should avoid emphasizing speed, which can disadvantage students with dyscalculia. When chosen carefully, games become the learning experience.

Chances are the curriculum you use already includes well-tested games, making it easier to find and select activities that target specific skills without having to create everything from scratch. Collaborating with teachers from the grade level above or below can also help you choose games students are already familiar with, while differentiating for the needs of your current learners. This both streamlines planning and allows students to focus on the math instead of learning a brand-new set of rules.

From Strategies to Planning: Making the Curriculum Work for Your Students

Up to this point, we've explored six of the most evidence-based instructional strategies for supporting students with dyscalculia: mathematical language, manipulatives, worked examples, think alouds, targeted feedback, and games. Knowing what works is important, but it's only half of the picture. The next step is learning how to weave these strategies into the lessons you already teach while keeping in mind the needs of your students.

Most of us don't have the luxury, or the time, to build our math curriculum from scratch. Instead, we inherit a curriculum that comes prepackaged, prescheduled and, many times, a district that insists we follow it "with fidelity". We don't get to choose the topics, the sequence, or maybe even the daily activities. We are told that if we just trust the program as written, the students will learn.

But we know that's not the whole story. We teach people, not programs. Thoughtful adaptations, based on leading cognitive science, let us meet the needs of our students while still honoring the intent of the program.

Let's see what this looks like in real life. We'll head back to Leo and take a peek at one of his upcoming place value lessons. We'll keep the curriculum exactly as written to start, then look for natural places to weave in the six strategies, making sure the lesson works for Leo, and his classmates, without throwing the pacing or structure off track.

The Who: <u>Leo</u>

As we have discovered, Leo is a second-grade student whose strengths shine in conceptual understanding and adaptive reasoning. He makes sense of new ideas by linking them to what he already knows and is quick to explain his thinking out loud. He's open to revising his ideas when presented with new information, and he approaches challenges with persistence, often surprising others with the creative connections he makes.

Leo's learning profile reflects elements of both the core number and memory subtypes of dyscalculia. Early in the year, counting and representing quantities accurately was a challenge, but his persistence is paying off. He can now count objects, at least up to 24, with careful one-to-one correspondence. When adding, he often counts on his fingers or draws detailed pictures that mirror the strategies he uses with manipulatives. These concrete tools are still his go-to for solving problems, especially when facts don't yet come automatically. Multi-step problems can feel overwhelming as he works to remember each stage, but his determination keeps him in the game, even when the process is slow.

Despite these challenges, Leo thrives when learning feels meaningful, interactive, and connected to his world. He's energized by hands-on experiences, visual supports, and opportunities to share his thinking with others. Low-pressure, high-engagement activities bring out his confidence, while timely, encouraging feedback helps him take risks and trust his own reasoning.

The What: <u>Place Value</u>

The following second grade lesson is based on activities commonly found in many elementary math curricula. Whether the context is stickers, base-ten blocks, or another familiar item, these lessons typically focus on representing three-digit numbers in multiple forms, comparing them by place value, and making connections between models, words, and equations. While the specific examples here are adapted from one curriculum, the structure, pacing, and mathematical goals will feel familiar to most teachers. This makes it a useful template for seeing how the six strategies can be woven into lessons you already teach, without replacing the curriculum you've been given. Together we'll look at how this lesson can be modified step by step, so the strategies feel practical and easy to apply in your own classroom.

Title: Collections of Stickers Grade 2 Unit 3 Lesson 3 Lesson Objectives:

- Use a place value model to represent three-digit numbers as hundreds, tens, and ones.
- Compare three-digit numbers by comparing the same place values (hundreds with hundreds, tens with tens, ones with ones).
- Read and write three-digit numbers in both standard and word form.
- Represent three-digit numbers in expanded form.

Lesson:

Counting Collections of Stickers

Begin with a simple context. "Aya bought 5 strips of ten stickers and 3 single stickers. How many stickers does she have?" Invite students to describe their thinking. Record the total as both a number word ("fifty-three") and an equation ($50 + 3 = 53$). Briefly connect these representations to a visual model of the stickers (five vertical rectangles and three small squares).

Next, introduce hundreds. "Stefanie then bought 1 sheet of 100 stickers, 3 strips of ten, and 6 singles." Ask a student to display this collection with place value materials (e.g., flats for hundreds, rods for tens, units for ones). Have pairs talk about how to represent this amount as an equation, then invite volunteers to share and explain.

Connect each part of the equation to the physical model. Record the number in words, standard form, and expanded form.

<u>Comparing Quantities</u>
Present a second collection: "Hassan has 1 sheet of 100 stickers, 6 strips of ten, and 4 singles." Have students represent this using place value materials and notation. Ask, "Who has more stickers, Stefanie or Hassan? How do you know?" Encourage multiple strategies—comparing place values, using a number chart, or reasoning with the models. Record both comparison statements (164 > 136 and 136 < 164).

<u>Extending The Numbers</u>
Pose a new question: "If Stefanie gets another sheet of 100 stickers, how many does she have now?" Record the new total (236) in standard, expanded, and word form. Ask students to identify what changed in the hundreds, tens, and ones places. "The hundreds place increased by one because she added another group of one hundred."

<u>Independent Practice</u>
 Have students work through similar problems in their math journals or activity pages. Each problem should involve:
- A visual model or "sticker notation."
- Writing the number in standard form, word form, and expanded form.
- Comparing two collections using >, <, and =.

<u>Discussion and Reflection</u>
Gather students to share solutions. Highlight connections between the physical models, sticker notation, number words, and equations. Emphasize that the hundreds place shows the number of hundreds, the tens place shows the number of tens, and the ones place shows the number of ones.

The How: <u>Putting The Profile to Use</u>

Before we change anything about this lesson, we need to pause and think about Leo. The first step is to look closely at the lesson and ask ourselves: "Where might he get stuck?" His learning profile gives us some clues.

From Struggle to Strategy

Once we know where the bumps might be, we can connect each challenge to a strategy that makes the path smoother. For Leo, who we identified as having difficulty with the core number and memory subtypes, holding several steps in his head at once can be tough. To help, we might break a task into smaller parts, give him a visual checklist, or tweak student pages so the steps are easier to track. Switching between representations can also be a sticking point. To support him, we can lean on the CRA sequence—starting concrete, then moving to drawings, and finally (maybe) to symbols. We can keep manipulatives consistent across the unit, show him exactly how to sketch what he's using, and give him blank number sentences to guide the transition to abstract work. With a little foresight, we're not scrambling in the moment—we're anticipating and planning for his needs.

From Strength to Strategy

Just as important as spotting the tricky parts is noticing what Leo does well. His strengths that we identified from the Stands of Mathematical Proficiency give us a place to start. We know he's strong in conceptual understanding and adaptive reasoning: he explains his thinking clearly, makes connections across ideas, and sticks with problems even when they're tough. We can build on these by choosing strategies that let those strengths shine. For instance, worked examples that highlight number relationships can validate his conceptual understanding. Think-alouds and targeted feedback can draw out his reasoning and encourage him to compare or revise his approaches. There's no single recipe here. Adapting a lesson isn't about changing the numbers or giving fewer problems to solve. It's about letting the student's needs and strengths guide your choices. And remember, even one small change is still meaningful progress. By tying each strength to one or more of the six strategies, we design lessons that not only address Leo's challenges but also celebrate his abilities.

So, let's take this lesson one step at a time and see how we can shape it to build on Leo's strengths while supporting his challenges. We'll focus on just one or two changes, knowing that you might make different choices, or see different opportunities. And that's great, as long as the choices we are making are grounded in what we know works best for students with dyscalculia.

Table 5.1: *Using Struggles and Strengths to Inform Strategy*

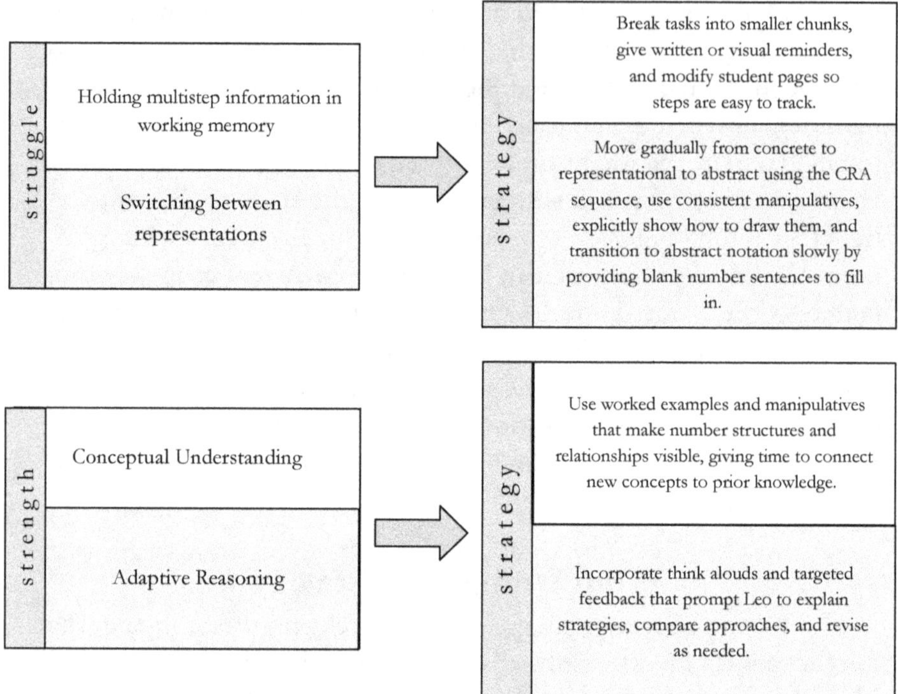

This chart helps us organize what we've thought about, but it's not always realistic to address every point at once. In a real classroom, we often have to choose the one or two strategies with the most leverage—especially when we're balancing the needs of many different learners at the same time. In this case, manipulatives stand out as a must-have for Leo, so we'll focus on making sure they're central to the lesson. That might mean giving him his own set to follow along with as we model, or, if that's likely to be distracting, inviting him to act as a "materials helper" by building the numbers for the class with the blocks.

Shifting the main activity to be partner-based is another easy adaptation that builds in peer feedback without adding any extra steps. Providing sentence stems or questions to ask each other can support more focused conversations, such as "First, I...", "I think this because...", or "Can you explain why...?". These prompts keep the dialogue centered on reasoning rather than just answers, and they help ensure that both partners are actively engaged in mathematical thinking.

Finally, keeping all modeled examples visible on the board throughout work time allows those examples to double as worked examples—a ready-made support for Leo that doesn't require changing the core structure of the lesson at all.

Reflection Activities for Teams and Book Study Groups Practice With Sofia

Profile: Sofia is our fifth-grade student who experiences *challenges* tied to the memory and reasoning subtypes of dyscalculia. She struggles with remembering multi-step procedures, recalling facts, and connecting known concepts to unfamiliar problems. When working independently, she prefers not to use manipulatives. As a result, she struggles to represent her thinking in pictures and numbers on the page.

At the same time, Sofia has considerable *strengths*. She demonstrates strong conceptual understanding, strategic competence, and a productive disposition. She understands how to use concepts, identifies important information, and sees herself as capable in math. Let these strengths guide your planning so your adaptations meet her needs while giving her opportunities to shine.

Your Turn:
Review Sofia's learning profile and identify:

- Which of the six strategies - mathematical language, manipulatives, worked examples, think aloud, targeted feedback, and games - could best address her struggles? (Struggle to Strategy)
- How might you connect those strategies to her strengths so she has opportunities to succeed and feel confident? (Strength to Strategy)

Then read the fifth-grade fraction lesson provided and identify:

- Where Sofia is likely to get stuck based on her learning profile.

- How you may adapt the lesson to support Sofia's success toward her goals.

Sofia's Lesson Adaptation Planning Chart

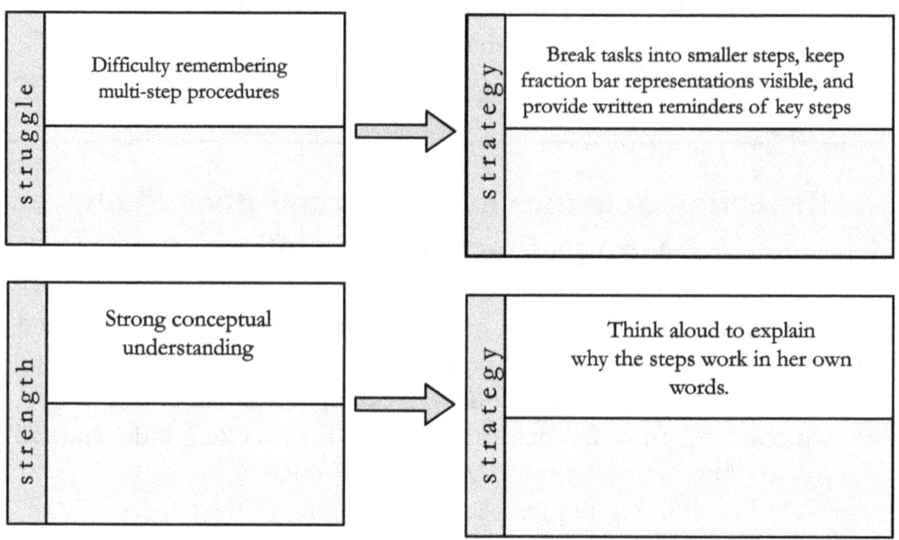

Tips for Use:
- You don't need to fill in every box. Choose the strategies with the most leverage for Sofia's learning.
- Think about overlaps where a single adaptation could address multiple needs.
- Keep her strengths in mind as much as her struggles. Both sides should inform the lesson plan.

<u>Compare Fractions</u>

Who has gone the farthest distance at the end of Day 1? Who has traveled the shortest distance?

How do you know? Turn and talk to a partner about how to compare these fractions.

(Students should recognize that 3/10 is larger than 1/6 and 1/8, so Lila drove the farthest. Diego drove the shortest amount because 1/8 is the smallest fraction.)

<u>Find Exact Distances</u>

Now let's figure out how many miles Lila actually drove. If the whole drive is 360 miles, what is 3/10 of 360? How can we find out?

Turn and talk: How would you solve this problem?

Use fraction bars to represent the whole drive (360 miles) and shade in 3/10 to see Lila's progress. Share solutions and explain your thinking.

Write The Equation

We found that 3/10 of 360 is 108 miles. How can we write this as an equation? $3/10 \times 360 = 108$

Independent Practice

Let's imagine it's now the end of Day 2. Each driver has made more progress on the drive:

- Amira: 3/6 of the drive

- Diego: 6/8 of the drive

- Lila: 7/10 of the drive

Answer the following questions using drawings and number sentences.

- Who has driven the most now?

- How do you know?

- How many miles has each person driven? Write an equation to show your reasoning.

- How does your equation connect to the fraction bar representation?

Discussion and Reflection

Gather students to share solutions. Focus discussion around writing multiplication equations for multiplying a fraction and a whole number.

References

Bird, R. (2021). *The Dyscalculia Toolkit: Supporting Learning Difficulties in Maths*. Corwin UK.

Fastame, M. C. (2021). Intervention programmes for students with dyscalculia: Living with the condition. In D. Lucangeli (Ed.), *Understanding dyscalculia: A guide to symptoms, management and treatment* (pp. 41–64). Routledge/Taylor & Francis Group. https://doi.org/10.4324/9780429423581-4

National Research Council. (2001). *Adding it up: Helping children learn mathematics* (J. Kilpatrick, J. Swafford, & B. Findell, Eds.). National

Academy Press. https://doi.org/10.17226/9822 Pennsylvania Training and Technical Assistance Network. (n.d.). *Concrete-representational-abstract: Instructional sequence for mathematics* [PDF]. PaTTAN. https://www.pattan.net/getmedia/9059e5f07edc-4391-8c8e-ebaf8c3c95d6/CRA_Methods0117

U.S. Department of Education, Institute of Education Sciences, What Works Clearinghouse. (2021). *Assisting students struggling with mathematics: Intervention in the elementary grades* (WWC 2021003). National Center for Education Evaluation and Regional Assistance. https://ies.ed.gov/ncee/wwc/PracticeGuide/21

Wall, H. (2022). *Teaching Students with Dyscalculia.* PolyMath Publishing.

Chapter 6
There's No I In Team: Family-School Partnerships

When we fail to connect, misunderstandings grow, and children fall through the cracks.

I remember sitting across from a mother during a parent–teacher conference. She told me she thought her daughter might have a learning disability in math. At the time, I shrugged it off. I had never heard of such a thing. Learning disabilities in reading, of course, but math? Her daughter could benefit from a tutor, I thought. She just needed to practice more and eventually things would click. They didn't; her daughter repeated the grade the following year. That moment has stayed with me, because it showed how limited my understanding was at the time, how unsupportive I was of the concerns of a parent, and how much my assumptions cost that student. Years later, I found myself on the other side of the conference table, this time as the mother of a child with dyscalculia. I've sat through countless meetings - some overwhelming filled with plans that never seemed to help, and others misleading, with teachers emphasizing only strengths so I walked away thinking she was finally on grade level. Even with my experience as a teacher, I often felt confused, uncertain, and afraid that I was failing my daughter.

What I've learned from sitting on both sides of that table is simple, that families and teachers want the same thing: for students to thrive. When we fail to connect, misunderstandings grow, and children fall through the cracks. When we work together, that partnership becomes one of the most powerful supports a student will ever have, and every child deserves that.

Setting the Stage: This Isn't About Newsletters

Too often, when educators talk about "family involvement," the conversation turns toward surface-level outreach: making colorful newsletters, sending home weekly emails, or hosting a family math night. Those things can be helpful, but they don't get to the heart of what really matters for students who struggle year after year.

This chapter is not a menu of activities. It's about **relationships rooted in shared understanding**. It's about seeing families not as recipients of information, but as partners who bring knowledge, history, and insight that no standardized test or report card can capture. It's about recognizing the emotional weight that comes with raising, or being, a child who struggles in math and building bridges of trust that allow true collaboration to happen. When families and teachers meet in this deeper way, communication goes beyond updates and checklists. It becomes a space where:

- Teachers acknowledge parents' lived experiences, including their own math struggles.

- Families feel heard, even when they bring hard questions or painful stories.

- Students see the adults in their lives working together, not in parallel, but in partnership. That's the work this chapter is about.

Why Families Matter

Families hold unique knowledge and advocate when systems fall short.

Parents and caregivers have the deepest, most continuous perspective on their child. They've seen how their child interacts with math outside the classroom, whether that's counting change at the store, measuring for a recipe, or playing games. They see their progress, their struggles, and their strengths. This insight can reveal patterns teachers don't always see in school, especially around persistence, anxiety, or the strategies a child invents at home. Families are also the ones advocating for assessments, accommodations, or outside support when schools don't move quickly enough. A strong relationship with families ensures that advocacy is informed, collaborative, and focused on solutions rather than conflict.

Families provide consistency and shape identity.
Consistency is everything for a child with dyscalculia. If the language or strategies change between school and home, confusion grows. Families help by reinforcing classroom strategies at home, but this communication must be two-way. Teachers also need to know what's working at home so instruction can align. That consistency doesn't just help with accuracy. It shapes how students *feel* about themselves as learners. A child who sees adults united in their support is more likely to build a positive math identity, rather than internalizing shame or believing they are "bad at math."

Families carry history.
Learning disabilities tend to run in families. That means when you sit down at the conference table, you may not only be facing a child's math struggles, but also the echoes of a parent's, and even grandparent's, own painful experiences. For some families and caregivers, math brings memories of frustration, shame, or even trauma. Recognizing this history is critical. When educators acknowledge the whole story, families feel seen and understood, creating the trust needed to move forward together.

Building Trust in Conversations

Trust isn't built in a single parent–teacher conference. It develops over time, through small, consistent interactions where families feel both seen and respected. For families of children with dyscalculia, conversations with educators can carry extra weight. These meetings may bring up years of frustration, discouragement, or even trauma around math learning, not just for the child, but for the parent, too.

Building trust also means respecting that families may take different paths. Some may rush to hire tutors, arrange private evaluations, or seek outside services. Others may place their faith in the school and expect more intensive support within the classroom. Still others may not be ready to act at all, because they are still processing what it means to have a child who struggles so deeply with math.

As educators, we may not always agree with the choices families make, but it's essential to remember that each decision is rooted in love and in the belief that they are doing what's best for their child. Respecting those choices, even when they diverge from our own professional instincts, is part of building trust.

Paolo Tan, in *Humanizing Disability in Mathematics Education*, calls this stance "cultural humility"— "a richly caring attitude that incorporates a lifelong commitment to self-evaluation and self-critique, to redressing power imbalances." Practicing cultural humility means acknowledging that families bring their own histories, hopes, and fears to the table. It means pausing to check our assumptions, recognizing the uneven power dynamics at play, and treating parents not as adversaries or passive recipients of expertise, but as essential partners.

Even if a family is not ready to take immediate action, communication is still a crucial step. By maintaining open, honest, and compassionate dialogue, we not only fulfill our role as educators, but also support families in their ongoing journey of understanding and acceptance. Sometimes, the most important thing we can offer is not a solution, but steady reassurance that we are walking alongside them.

From Trust to Talk: Setting the Stage

Although trust is built in small steps, it's sustained in how we actually hold conversations. Whether a parent has requested the meeting or the teacher has, those first moments matter. It's tempting to jump straight into test scores, lesson plans, or solutions. After all, time is short, and families often look to us for answers. But lasting partnership doesn't begin with advice; it begins with listening.

At the same time, families look to teachers for clarity, and conversations need structure to avoid drifting into either vague reassurances or overwhelming detail. One simple framework that supports trust is to shape every conference around three parts:

1. **Here is what I observe** – facts and patterns from the classroom, with space for family input.
2. **Here is why I think it is happening** – your professional reasoning, with space for family input.
3. **Here is what we can do** – next steps, with space for family input.

Listen and Validate - Even When You Speak First

Conversations with families are shaped as much by tone and approach as by content. The way we open a conversation often matters more than the words themselves. Sometimes families call a meeting ready to share, and sometimes teachers have to open the conversation. Either way, listening and validating are the cornerstones.

- **Start with acknowledgment.** Begin with a neutral observation and an open invitation. This sets the tone and gives families the chance to guide the direction of the conversation. For example: *"I've noticed some ups and downs in Sofia's work in our unit of multiplying fractions. Before I share more, I'd love to hear how math has been feeling at home."* This gives parents space to share first, while signaling genuine curiosity about their perspective.

- **Lead with strengths.** As emphasized in Chapter 3, highlighting strengths isn't about glossing over difficulties. It's about building an honest, balanced picture. Put the strength first in your examples to communicate hope and possibility. For example: *"Leo is great at working with pictures. He can look at a diagram and explain it clearly. What's harder for him is matching them to the written math problems. I think we can use his strength with pictures to help him make sense of the numbers."* This framing reassures families that their child is more than their challenges.

- **Validate concerns.** When parents share their frustrations, resist the urge to move straight into fixes. A simple response like, *"I can hear how hard that has been for you both,"* or *"It sounds like homework time has been really stressful for both of you. You're right. We need to find a way to make that feel less overwhelming,"* acknowledges the emotional weight behind their words. Validation communicates respect and empathy—two essentials for building trust.

Bridging The Language Gap

The second step, "This is why I think this is happening"—is often where well-meaning teachers lose families. In our effort to be precise, we sometimes slip into jargon or overly technical explanations. Parents may hear terms like *processing deficit* or *executive function challenges* and leave the meeting more confused than informed. What was meant to clarify instead creates distance, and families walk away with a list of problems but nothing constructive to hold onto.

Bridging the language gap means translating professional insight into clear, compassionate language. Instead of saying, *"Leo's difficulties with working memory affect his sequential processing,"* you might say, *"Leo understands the math concepts, but he loses track of the steps when there are too many to hold in his head all at once."* Both are accurate, but only one feels accessible.

Whenever possible, ground your explanations in everyday examples families already know. Comparing math steps to following a recipe, remembering the rules of a board game, or keeping track of items while grocery shopping helps parents connect abstract difficulties to real-life situations. These analogies not only build understanding but also give families language they can use with their child that feels supportive rather than clinical.

Another common language gap arises in how we talk, or don't talk, about assessments. As teachers, we know the difference between a screener, a standardized test, a formative task, and an end-of-unit exam. Parents usually don't. And why would they? Without clarity, a single low score, or a folder full of them, can feel catastrophic. Taking time to explain, *"This screener is just a snapshot of how Sofia handled math fluency today under timed conditions. It's not a judgment of her ability, but a tool to guide our instruction,"* can reduce anxiety and foster trust.

Ultimately, bridging the language gap isn't about watering down our professional knowledge. It's about making sure families leave the room with a clear, accurate picture of their child, and the confidence that they are truly partnered with us in moving forward.

Making Assessments Clear

Assessments. To families they all look the same: like tests. To teachers, the difference between a screener, a unit test, a formative check, and a standardized exam is second nature. Without clarity, especially *before* results are sent home, parents may walk away thinking their child is "failing everything" or, on the other hand, "doing just fine," when the reality is much more nuanced.

A helpful way to bridge this gap is to briefly name the purpose behind each type of assessment when sharing results:

- **Screeners:** *"These are short, timed assessments that we do a few times a year without any accommodations. They provide a quick snapshot of where a student might need more support in class. Think of them more like getting your temperature taken at the doctor than a report card."*

- **Formative Tasks:** *"These are assignments done in class after one or a series of lessons. They help me see what concepts are clear and what might need more practice. They're not about grades as much as guiding my teaching."*

- **Unit Tests:** *"These happen when we finish a unit of study. They show what your child learned from that unit specifically in relation to grade level standards."*

Standardized Tests: *These are given across a district, or state, and compare your child's performance to all the standards for their grade level. They're helpful for seeing overall trends, but they don't capture capture everything your child knows, especially creativity, perseverance, or how to approach real world problems."*

A Note of Caution: Each school handles assessments differently. Some include homework in grading. Others may attach grades to formative tasks, while some do not.

Practical Routines That Build Collaboration

Parents should never leave a meeting feeling like the burden of *fixing everything* has been shifted onto them

The third step, *"Here is what we can do"*, works best when it begins with the school's plan, not a list of things families should take on at home. Parents should never leave a meeting feeling like the burden of fixing everything has been shifted onto them. That's not partnership, it's pressure. By first communicating, *"Here is what we can do in school"*, teachers demonstrate professionalism and clarity.

From there, invite parents' feedback: *Does this feel right to you? Have you seen success with anything similar at home?* If families are open to it, you can then explore light, realistic ways to reinforce learning outside of school. But it's equally important to respect that not all families will be in a position to do extra work at home for reasons ranging from time constraints, to language barriers, to the possibility that a parent or sibling may also have a disability that makes additional support difficult.

Naming what the school will do first, and then extending an invitation for home involvement, makes collaboration feel like a choice rather than an obligation.

As both a teacher and a parent, I've experienced this tension firsthand. When my own daughter was struggling in school, my family chose not to pursue tutoring or intensive at-home support right away. Adding more at home felt premature, and we didn't want extra work to blur the progress she was—or wasn't—making from school-based supports. That experience reminded me how vital it is for schools to lead with, *"Here is what we will do,"* and then invite families into the conversation about what's realistic at home.

That's why practical routines matter. Families feel more confident when they see that the school already has a clear plan in place—something structured, repeatable, and transparent. One tool I've adapted from Julia Aguirre and colleagues in *The Impact of Identity in K–8 Mathematics* is a simple conference template. It offers a consistent way to share strengths, challenges, and next steps, and it can be revisited and revised across meetings, so families see growth over time.

Parents often leave meetings with pages of notes but little clarity about what comes next. A tool like this helps reduce that overwhelm. Instead of families trying to remember every detail, they leave with a clear snapshot of what was discussed and a shared plan to guide them forward. This adapted framework becomes a powerful communication tool that sets the stage for true collaboration—and it also doubles as helpful documentation when reaching out to school-based support teams for additional services.

Here's how this template, found in Appendix D, might look in practice using Leo as an example.

Table 6.1: Leo's Conference Table: Before The Conference Begins

Area	Teacher Action	Parent Action	Student Action
Strength: Leo explains his thinking out loud, which shows strong reasoning.	Provide structured opportunities for Leo to share strategies with a partner or small group.		
Area For Improvement: Loses track of multi-step problems.	Break problems into smaller steps and use a checklist for solving.		
Area For Improvement: Addition fact fluency	Give opportunities to practice doubles and near doubles with math games outside of math instruction time.		

Sample Script for Using the Template

"I've filled in a few notes ahead of time about what I've noticed and how I plan to support Leo in class. The rest we can fill in together. Some boxes may

stay blank—and that's fine. The goal is to make sure we're all clear on what's happening in school and what's realistic at home. I'll send a copy of this home with you today, and next time we meet, we can look back at it together and see what's been working and what might need to change."

This format makes clear what *you*, as the teacher, are committing to in the classroom first. It then leaves room for parents and students to contribute, but without pressure to "fix" things at home. Over time, the chart itself becomes a living record of growth and collaboration—something both families and teachers can return to as evidence of progress and next steps.

Table 6.2: Leo's Conference Table: After The Conference

Area	Teacher Action	Parent Action	Student Action
Strength: Leo explains his thinking out loud, which shows strong reasoning.	Provide structured opportunities for Leo to share strategies with a partner or small group.		Speak out loud about strategies when solving problems.
Area For Improvement: Loses track of multi-step problems.	Break problems into smaller steps and use a checklist for solving.	For now, just wait and see how this works at school before changing anything at home.	Try using the checklist when working through problems.
Area For Improvement: Addition fact fluency	Give opportunities to practice doubles and near doubles with math games outside of math instruction time with a friend.	If time allows, play a doubles game at home once a week.	Practice doubles games in class.

Being Transparent About Roles—
and Knowing When to Call in the Team

Families deserve clarity about who does what when a child is struggling. Without that transparency, they may leave meetings expecting daily tutoring from the teacher or assume that nothing more can be done if classroom strategies aren't working. Both misunderstandings lead to frustration.

That's why it helps to be upfront about the teacher's role. Share what you can realistically do in your classroom—small-group instruction, targeted practice, regular check-ins and also what falls outside your scope, such as one-on-one daily tutoring or specialized testing. Framing it this way communicates professionalism: *I am committed to supporting your child with the tools I have, and if more is needed, I will make sure the right people are involved.*

This is where the support team comes in. Even the most dedicated classroom teacher cannot, and should not, do it all alone. If a student shows minimal progress after a cycle of classroom interventions, that's when it's time to raise the concern with the school's support team. These teams, whether through your school's intervention team (called MTSS or RTI in my district), child study, or special education pathways, look at multiple sources of data, consider the whole child, and collaborate with the family to decide next steps.

It's important to frame this step not as a failure, but as a mark of thoroughness and care. Families should hear: *"If we reach the point where classroom strategies aren't enough, with your permission, I'll bring this to our team. Together, we'll review what's been tried, what the data shows, and what supports might help. You'll be part of every step."*

By being transparent about your role, and equally clear about when to bring in additional support, you prevent false expectations. Families leave knowing what you're committing to now, what comes next if progress stalls, and that they'll never be left out of the process.

Moving Forward with the Team

For classroom teachers, the most important contribution is the documentation and observations you've already been gathering. The strategies you've tried, the routines you've adapted, and the student responses you've tracked become the foundation of the team's work. Instead of starting from scratch or repeating old interventions, the team can build on what's been done, making decisions with momentum rather than hesitation.

This process also protects families. Too often, parents are asked to retell the same concerns or defend why their child needs help. When you arrive at the table with clear, organized notes, it shifts the dynamic. The narrative is no longer "the student is struggling," but rather, "here's what we've tried, here's what we've noticed, and here's what we still need." That changes the conversation from vague worry to advocacy and actionable planning.

As a classroom teacher, you don't need to have all the answers. What you need is to remember the importance of your role:

- **You are the first responder**—the one who notices, documents, and initiates support.

- **You are the bridge**—connecting the lived classroom experience with the expertise of specialists.

- **You are the advocate**—ensuring the student's strengths, challenges, and history are represented accurately.

When the team meets, your voice matters because you bring the day-to-day lens. And by sharing your records—the error pattern checklist, strands of mathematical proficiency, CRA goals, and conference collaboration plans—you give the team a realistic picture of the student's learning journey. That level of transparency helps the whole group design next steps that are forwardmoving, not backward-looking.

In the end, moving forward with the team isn't about handing the student off. It's about widening the circle of support, making sure the student's story is understood, and keeping collaboration at the center of the process. To help you process these ideas and connect them to your own practice, the following reflection activities offer space to step back, discuss, and plan for action.

Reflection Activity For Teams And Book Study Groups

Chapter Specific Reflections

1. Think about a recent family meeting or parent-teacher conference. What were some ways you made clear both your role and what the school could provide?

2. Review your own record keeping practices. Do you have a system, like the conference table, that could be shared with a team to show what's been tried? If not, what might you adapt from the tools in the book?

3. Consider a student you are currently concerned about. If progress stalls, what would "bringing it to the team" look like in your school? Who would be involved and what information would you want to bring forward?

Whole Book Reflections

1. Looking back over the book, which ideas or tools feel most doable in your current classroom practice? Which will require more collaboration or support to implement?

2. How has your perspective on dyscalculia shifted from deficit based to strengths based?

3. Can you name specific ways you'll communicate student strengths to families and teams?

4. Revisit the goals stated in the introduction: building your confidence, seeing students' gifts, and keeping teaching manageable. Where have you noticed growth in yourself around these goals? Where do you still want to grow?

5. Consider your role within your professional community. How might you use what you've learned here to support colleagues, advocate for students, or shape school-wide practices?

References

Aguirre, J., Mayfield-Ingram, K., & Martin, D. (2013). *The Impact of Identity in K-8 Mathematics: Rethinking Equity-Based Practices*. National Council of Teachers of Mathematics.

Tan, P. (2019). *Humanizing Disability in Mathematics Education: Forging New Paths*. NCTM.

Epilogue

At the heart of this book is a simple conviction: every student, no matter how they process numbers, deserves to know they belong in the math classroom. Belonging is not just about access to curriculum. It's about being seen, valued, and trusted as a thinker.

This conviction is not theoretical for me. It is personal. Watching my daughter navigate school reminded me daily that the math classroom can be both a place of joy and a place of alienation. It pushed me to look closer at my own teaching, to ask harder questions, and to create more intentional spaces where students who struggle with math can still feel like mathematicians. Olivia's journey has grounded this work in real stakes: this is about children, our children, finding their voices in math.

When I began writing, I named three goals: for you, for your students, and for myself. In Chapters 1 and 2, my hope was that you'd feel more confident and informed about dyscalculia. You didn't need to throw out your curriculum or reinvent yourself as a teacher. What you needed were clearer ways to recognize patterns, name what you see, and respond with intention. My hope is that you left those early chapters not burdened but reassured: you already have many of the tools you need, and with sharper language and focus, you can use them even more effectively. In Chapters 3 and 4, we shifted the focus to the children themselves. Too often, students with dyscalculia are seen only through the lens of deficit. I wanted to turn that story around. These chapters invited you to notice and celebrate what students bring: their reasoning, persistence, and creativity. By anchoring instruction in strengths and by using structures like CRA to meet them where they are, we can help students develop positive math identities and hear their own voices in the classroom.

Chapters 5 and 6 gave me space to reflect on my own journey as both teacher and parent. Writing about planning instruction and building partnerships with families helped me gather years of missteps and insights into something more coherent. What I know now is this: teachers don't need another list of impossible demands. We need clarity, purpose, and practices we can actually carry into our classrooms. Pulling these chapters together has been a way of distilling what has mattered most in my work and offering it forward to you.

One theme has surfaced again and again: no teacher can do this alone. Nor should we. Our role is essential, but it is not solitary. We are the first responders, the ones who notice when a student is stuck. We are the bridges, connecting what we see daily to what specialists and families know from other contexts. We are the advocates, ensuring that a child's strengths are never lost in the conversation about their needs.

That's why we build partnerships with families, colleagues, and support teams. Each chapter of this book has emphasized collaboration, because sustainable change comes from widening the circle. Students thrive when the adults in their lives share a plan, a language, and a commitment to seeing them succeed.

The work is not easy. Some days you will feel the weight of responsibility pressing hard. Some days progress will come slower than you want. But small steps add up. Every time you pause to notice a strength, to name a success, to invite a parent's voice into the process, you are reshaping what it means for a student with dyscalculia to learn math.

When we choose to see our students' strengths, when we commit to collaboration, and when we lead with both professionalism and care, we're doing more than teaching math, we're shaping identities. We are showing children that they are capable learners, that their voices matter, and that their stories belong in the classroom.

That work is important. That work is lasting. And the good news is, you don't have to do it alone. For my daughter, and for every child who walks into our classrooms, that makes all the difference.

Appendix A Student Error Pattern Checklist

Student Name: _________________________________ Date: ___________

Grade/Class: _________________________________

Directions: Check all that apply as you review student work, observe during instruction, or reflect on student responses. Look for clusters. They may point to where breakdowns are happening most consistently.

Core Number

- ☐ Counts slowly or loses track when counting.
- ☐ Struggles with subtilizing (recognizing small amounts without counting).
- ☐ Confuses similar numbers (e.g. 13 and 31, 42 and 24, 101 and 110).
- ☐ Difficulty with hierarchical inclusion (understanding that a number is made up of smaller numbers).
- ☐ Struggles to count backward or start at a number that is not zero or one.
- ☐ Struggles to represent quantity in different ways (numbers, pictures, and/or words)
- ☐ Difficulty rounding or estimating.
- ☐ Difficulty with place value understanding.

Upper Elementary & Middle School

- ☐ Struggles to demonstrate a conceptual understanding of rational numbers (whole numbers, fractions, and decimals).
- ☐ Has limited ways to represent rational numbers (whole numbers, fractions, and decimals).
- ☐ Unable to use multiple representations of rational numbers (whole numbers, fractions, and decimals) to solve problems.
- ☐ Difficulty understanding how operations (addition, subtraction, multiplication, division) effect rational numbers (whole numbers, fractions, and decimals).
- ☐ Understands the equal sign as giving and answer as opposed to a symbol that shows an equivalent relationship between terms.

____

____

Memory

- ☐ Uses fingers or manipulatives for simple calculations.
- ☐ Difficulty performing mental math.
- ☐ Forgets previously learned procedures.
- ☐ Difficulty recalling learning number facts.
- ☐ Difficulty remembering mathematical vocabulary.
- ☐ Struggles to move from concrete to abstract representations.
- ☐ Seems to get lost in the steps of a procedure.

____

____

Reasoning

- ☐ Difficulty explaining thinking.
- ☐ Struggles to make connections between concepts.
- ☐ Reliance on rote learning.
- ☐ Applies the same strategy regardless of problem type.
- ☐ Confuses steps in procedures.
- ☐ Difficulty decision making when problem solving.
- ☐ Confuses basic logic principles (If…then…).

____

____

Visual-Spatial

- ☐ Uses visuals ineffectively.
- ☐ Reverses numbers or symbols.
- ☐ Difficulty seeing patterns.
- ☐ Work appears disorganized and hard to read.
- ☐ Struggles to line numbers up by place value.
- ☐ Difficulty interpreting spatial organization of numbers (e.g. decimal notation, exponent).
- ☐ Difficulty creating, using or interpreting number lines.
- ☐ Difficulty creating, using or interpreting graphs.

____

____

Appendix B Strands of Mathematical Proficiency

Student Name: _________________________ Date: ______________

Grade/Class: ____________

Directions: Please check off all the behaviors that apply to the student.

Conceptual Understanding

- ☐ Understands why mathematical ideas are important.
- ☐ Understands the contexts in which a mathematical idea is used.
- ☐ Learns new ideas by connecting to what they already know.
- ☐ Explains the method they use to solve a problem.
- ☐ Represents mathematical situations in different ways based on different purposes.
- ☐ Connects what they know to new or unfamiliar problems.

Procedural Fluency

- ☐ Knows when to add.
- ☐ Has strategies for addition.
- ☐ Addition strategies are efficient.
- ☐ Addition strategies arrive at the correct answer.
- ☐ Knows when to subtract.
- ☐ Has strategies for subtraction.
- ☐ Subtraction strategies are efficient.
- ☐ Subtraction strategies arrive at the correct answer.

Grades 3 and Up

- ☐ Knows when to multiply.
- ☐ Has strategies for multiplication.
- ☐ Multiplication strategies are efficient.
- ☐ Multiplication strategies arrive at the correct answer.

☐ Knows when to divide.

☐ Has strategies for division.

☐ Division strategies are efficient.

☐ Division strategies arrive at the correct answer.

Strategic Competence (Problem Solving)

☐ Formulates a problem that requires math to solve.

☐ Demonstrates understanding of a problem's important information.

☐ Represents a word problem using numbers, pictures, words, diagrams, or graphs.

☐ Identifies which information is unimportant in a problem.

☐ Uses an approach to make sense of a new problem type (using manipulatives, creating drawings, etc.).

Adaptive Reasoning

☐ Identifies and explain patterns they see.

☐ Explains their reasoning for an answer.

☐ Determines whether or not a strategy will work.

☐ Abandons strategies that seem ineffective and generates alternative plans.

☐ Justifies whether or not an answer is reasonable.

Productive Disposition

☐ Sees themselves as capable of learning mathematics.

☐ Believes math can be useful.

☐ Shows perseverance.

☐ Has a growth mindset towards mathematics.

☐ Has a positive disposition towards mathematics.

Appendix C CRA Goal Setting Tool

Name: _______________________________

Date: _______________________________

Task:

	Strengths Highlighted	Possible Goals	Planning Implications
Concrete Stage			
Representational Stage			
Abstract Stage			

Appendix D Conference Collaboration Template

This template is designed to help teachers, families, and students build shared plans during conferences. Fill in what you can ahead of time, and complete the rest collaboratively. Some boxes may remain blank—and that's fine.

Date: _______________________

Student Name: ____

Family Names: ____

Teacher Names: ____

	Area	Teacher Action	Parent Action	Student Action
Strength				
Area For Improvement				
Area For Improvement